THE
UNIVERSE
EXPLAINED

Heather Couper and Nigel Henbest

A COSMIC Q&A

FIREFLY BOOKS

A Firefly Book

Published by Firefly Books Ltd. 2018

First printing

Library of Congress Control Number: 2018937766

Library and Archives Canada Cataloguing in Publication
Couper, Heather, author
  The universe explained : a cosmic Q&A / Heather Couper and Nigel Henbest.
Includes index.
ISBN 978-0-228-10082-9 (softcover)
  1. Astronomy--Popular works. 2. Astronomy--Miscellanea. 3. Cosmology--
Popular works. 4. Cosmology--Miscellanea. I. Henbest, Nigel, author II. Title.
QB44.2.C687 2018   520   C2018-901649-3

Published in the United States by
Firefly Books (U.S.) Inc.
P.O. Box 1338, Ellicott Station
Buffalo, New York 14205

Published in Canada by
Firefly Books Ltd.
50 Staples Avenue, Unit 1
Richmond Hill, Ontario L4B 0A7

Cover and interior design: Kimberley Young

Printed in China

The Milky Way Galaxy, as viewed from the Earth, Canyonlands National Park, Utah

CONTENTS

INTRODUCTION

This book is dedicated to you.

You, and all the thousands of people who come to our presentations eager to learn about the cosmos. From America to Australia, China to Colombia, the Q&A session of our lectures is often the best part.

The Q&As are very inspirational. They may not always relate directly to the presentation, but that's half the fun of it! We may have just given a talk about Mars when someone asks us "What's a black hole?" Once, after an overview of the history of astronomy, we were asked: "Can I see planets with the unaided eye?" The Q&A almost always turns into a fantastic audience discussion about astronomy. It's a great way to share our curiosity about the cosmos.

It doesn't stop at presentations either. At one of our local country pubs in England, a query about sky sights erupted into action when patrons poured out into the parking lot to spot the planets and watch the International Space Station cruise overhead.

It's plain to us that astronomy drives our imagination, and questions are the power that fuels the drive. One of the most fundamental questions is whether there's life somewhere other than Earth. We still don't know the answer. We're pretty up-to-speed on how the universe began — but why, and what caused it? Questions like these have driven the human quest to understand the universe and our place in it. If we hadn't started asking how the heavens worked, we'd still believe that we lived on a flat Earth with stars stuck to a dome above.

Now we know that Earth is a vibrant, living member of the Solar System, and we are sending space probes (and soon, people) to our fellow worlds. We've discovered that our Sun is just one of 200 billion stars in our Galaxy, and that there are billions of other galaxies in the universe.

It's *your* questions that have inspired this book. We had so many of them that a full third didn't fit, and so we've pared them down to just under 200.

There are some apparently simple questions: Why is the sky blue? What makes the Sun red at sunset? The answers, however, aren't straightforward. Then there are the hard ones: What's dark matter? Why do some stars explode?

We've had so much fun answering these questions and providing images to go with them. Like the pictures in our public presentations, many of the images included in this volume are seldom seen. We've included rare historical images and the very latest photos from the world's biggest space telescopes. When we couldn't find an image — how do you photograph alien life or a parallel universe? — we asked talented cartoonist and astronomer, Geoff Burt, to come up with the goods.

We've grouped related questions together, but you don't have to read it chapter-by-chapter — just dip in as you will.

For every question we've tried to answer, you'll find a dozen of your own. Enjoy, and follow your curiosity where it leads you!

Heather Couper and Nigel Henbest,
March 2018

Link to the cosmos: the giant Parkes radio telescope in Australia has tuned into astronauts on the Moon and distant galaxies — and it's also on the hunt for aliens . . .

SKY SIGHTS 1

'What are the northern and southern lights?

The northern and southern lights — the aurora borealis and the aurora australis — are the coolest sky sights you can see. The heavens glow with glorious swirling curtains of changing colors, and the dark skies are pierced with dramatic rays. It's no wonder that cruise ships and planes ferry tourists by the thousands toward Earth's poles to witness this beautiful celestial phenomenon.

Aberdeen, in the north of Scotland, has a nickname for the many aurorae visible there each year. They're affectionately known as "The Merry Dancers."

We don't have to look too far to discover the perpetrator of this cosmic light show. It's the Sun, which increases its magnetic activity roughly every eleven years (see Chapter 6 for more info). Our normally quiescent local star hurls fast-moving electrons and protons into the Solar System. Being electrically charged, these rush toward the Earth's magnetic poles as they pass our planet, lighting up the sky like gas in a neon tube.

The dazzling colors of the aurorae come from the elements in Earth's atmosphere zapped by the Sun's awesomely energetic particles. Oxygen atoms at the highest altitudes glow red; medium-level oxygen gives the aurora its characteristic green hue, while nitrogen provides the blue and purplish tinges.

Not close to a pole? No worries. If the Sun has a particularly violent magnetic storm, aurorae can be seen well outside the normal polar limits, as close as 20 degrees to the equator. In the Northern Hemisphere, we once saw an aurora near Oxford, in the middle of England; they have been seen in Virginia and have even been reported in the south of France. South of the equator, aurorae are common in New Zealand, and as far from the South Pole as Tasmania and South America.

Don't be too disappointed if your aurora tour doesn't come up with the goods. The Sun is notoriously unpredictable with its magnetic hic-cups. The best time to spot aurorae is just before and after sunspot maximum — check the Web before you go. If your light show doesn't show, there will still be some sensational dark night skies to gaze upon.

The northern lights, captured by astronaut Tim Peake on the International Space Station.

Why is the sky blue?

The first person to work all this out, in 1871, was the British physicist Lord Rayleigh — an unusual scientist who discovered the rare gas argon, and then moved from investigating the natural world to the supernatural when he became president of the Society of Psychical Research.

Rayleigh realized that the sight of a beautiful blue sky is all due to the tiny gas molecules making up air. As the Sun's light shines down through the atmosphere, it's passing countless oxygen and nitrogen molecules, each of which can bend its path very slightly. Sunlight is made of all the colors of the rainbow, and the gas molecules are better at deflecting the shorter wavelengths of blue light than the longer red wavelengths — an effect now called "Rayleigh scattering."

Some of the blue light is deflected so much that it gets scattered sideways, and from here it's batted all over the place by the gas atoms around us. As a result, this short wavelength light can reach our eyes from any direction, and we see a blue sky wherever we look.

The clear blue sky of Antarctica contrasts with the dazzling white snow.

Why does the Sun turn red at sunset? What's the Green Flash?

"Red sky at night, sailor's delight." That's one of many old country sayings, but — unlike many — it's actually true: red evening skies really do herald fine weather the next day.

To find out why, we must return to our question about why the sky is blue (see opposite). After the air molecules have batted away some of the Sun's blueness and scattered it over the sky, the light that comes through to our eyes from the Sun is richer in longer-wavelength red light. As the Sun sinks toward the horizon, its light has to travel through more and more of the Earth's atmosphere, and it appears redder and redder.

But why are some sunsets more colorful than others? After the Sun has gone down, its reddish rays may continue to shine on clouds high in the western sky. If there's bad weather over the horizon, sunlight struggles to get through, and the sunset looks pretty tame. On the other hand, if the air to the west is clear for hundreds of miles, the last rays of the Sun can play on the evening clouds in a fantastic display of red colors. In mid-northern latitudes, weather systems tend to move from west to east, so this region of fine weather will move in the next day — true to the ancient sailor's prediction.

As for the rare Green Flash: like a prism, the atmosphere splits up the light from the Sun, so that the red disc has a fringe of yellow and green light on top. Usually this rainbow border is drowned out by the brilliant red disc; but when the atmospheric conditions are exactly right (the air lies in stable layers), the green fringe is still shining over the horizon when the rest of the Sun sets, and we observe the Green Flash.

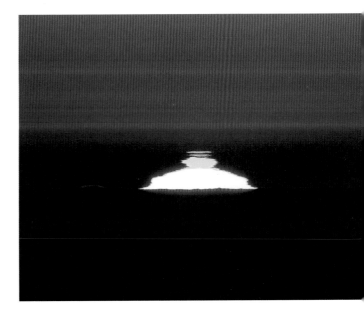

The top of the red setting Sun flashes green.

How many stars can you see?

Hang on, you say — I can only see a few stars from downtown, where I live. From Chicago, for instance, you'd be lucky to see 40 stars in the sky. But that's because of light pollution, which we'll be tackling later (see page 18).

No matter how many stars you can see, you'll notice that they aren't all of the same brightness. To categorize stars' brightness, astronomers have come up with a ranking system called magnitudes. Magnitudes are like golf handicaps: the lower, the better. Magnitude 1 stars are the brightest in the sky; they're 2.5 times brighter than magnitude 2 stars (like those in the Dipper), which are 2.5 times brighter than those of magnitude 3 . . . And so it goes on to magnitude 6, the dimmest stars that you can see — they're 100 times fainter than magnitude 1. Just to add to the confusion, some brilliant stars have *negative* magnitudes: Sirius, the brightest star in the sky, comes in at minus 1.46.

If you're hungry to see more stars, buy a pair of binoculars or a telescope (see Chapter 4). Even a small telescope can capture over a million stars — that should keep you going for a lifetime or two.

F ewer than you'd think! Ask people to guess, and they'll say "millions," but the actual answer is 9,096; and that's for the whole sky. On any particular night, only half of these are on show (as one hemisphere of the sky is hidden by the Earth beneath our feet), so you get to see around 4,500 — and that's on a perfect night with clear conditions.

This figure was arrived at by Dorrit Hoffleit, a respected astronomer at the Yale Observatory, who died just weeks after her 100th birthday in 2007. An expert on variable stars (stars that change in brightness), she compiled a large catalog of all the stars visible to the unaided eye. The Yale Bright Star Catalog was a phenomenal achievement.

There's a popular saying "the night has a thousand eyes," but out in the desert you'll find many more stars looking down on you.

Can I see the Moon, stars or planets during the day?

There's a long-standing story that you can see stars during the day if you stand at the bottom of a deep well, and the star lies in the tiny patch of sky that's visible directly overhead. But don't bother risking your neck — and possible drowning — to check this out. It's impossible anyway: the blue sky is too bright and washes out the faint star's light.

But the brighter planets are different. Venus is ten times brighter than any star, and we've seen it many times just before the Sun has set. If you have good eyesight, and the sky isn't hazy, you can pick out Jupiter as well. A bigger challenge is Mars: every two years, when it's closest to Earth, Mars outshines Jupiter — but the trouble is that the Red Planet is then low in the sky during daylight.

The best way to check out the planets in daylight is to wake up very early. Catch Venus, Jupiter or Mars glowing in the sky before dawn, then keep an eye on it as the sky brightens and the stars fade from view, and see if your chosen planet is still visible after the Sun has risen.

Or, use an astronomy app to find out when one of these planets is near the Moon. The Moon can act as your guide because it is visible during daylight.

During the daytime, the Moon is hidden from us only when it is Full; that's when the Moon is opposite to the Sun, so the Moon rises as the Sun sets. Throughout the rest of the month, you can see the Moon in the daytime sky: in the after-noons when it's not yet Full, and in the mornings after Full Moon.

Venus and the Moon, seen during the daytime

If you see clouds well after sunset — when the sky is nearly dark — your eyes aren't fooling you. Floating like rippled white sheets in the twilight, these are noctilucent clouds (night-shining clouds). You'll see them best from latitudes of 50–70 degrees north or south, when the summertime Sun illuminating them isn't far below the horizon.

Unlike conventional clouds — cumulus, stratus and nimbus — they're not made of water droplets. And while these ground-hugging clouds lie at altitudes of around 3,000 m, noctilucent clouds hover at the edge of space, reaching heights between 76,000 and 85,000 m.

Like cirrus clouds (6,000 m high), noctilucent clouds are composed of ice crystals. But ice crystals need a nucleus to form around, and here's where the astronomy kicks in. It's still a controversial idea, but the best bet is that the nuclei for noctilucent clouds — tiny particles of dust — are micrometeorites resulting from the break-up of larger meteoroids in space (see Chapter 8).

That's not the only theory. Reports of noctilucent clouds peaked in 1885, two years after the eruption of Krakatoa. Was volcanic dust to blame — or were more people gazing at the glorious sunsets caused by the explosion, and spotted the clouds afterward?

Night-shining clouds are called noctilucent clouds.

In the end, it could be largely because of us! Emissions from space shuttle exhausts have created mini-noctilucent clouds. Some scientists think it's more collateral damage from the greenhouse effect. Others reckon that the doubling of methane levels over the past 100 years (blame the cows!) has pumped more water vapor into the atmosphere, which can then freeze out at the edge of space.

Noctilucent clouds are still an enigma. But to see their eerie, bluish glow on a calm summer night is a glorious and mysterious sight.

Ghostly wisps of noctilucent clouds, seen here over Denmark in 2010, float almost at the edge of space.

Why do the stars look fainter than they used to?

A lady in her 80s came up to us after one of our presentations. "Have the stars faded?" she asked. "I can't see nearly as many as I could when I was a little girl."

No — her eyesight wasn't to blame. It's all our fault.

Since the Industrial Revolution, we have become adept at polluting our planet. And the skies are no exception. We're all familiar with the orange glow from sodium streetlights that warns you that a town is just over the horizon. Or the glare from sports lighting, which is seldom aligned to illuminate the field — a lot of it goes up into the sky. Then there's security lighting, which stuns you with a dazzling blast when you least expect it.

That's why the stars *seem* to have faded. They aren't actually any dimmer, but the glowing night sky makes it harder to see them.

Looking down from space, you can see the culprits — the countries, cities and towns that are producing the most waste light that streams upward. Check the image here from the Suomi-NPP satellite, and it's clear that light pollution is a particular blight on North America, Europe and eastern Asia — the most affluent and indus-trialized parts of the world.

From Switzerland, in the heart of light-polluted Europe, you'd have to travel 1,360 km to enjoy completely dark skies — to northwest Scotland, Algeria or Ukraine. From Chicago, the heart of America, truly black skies are 2460 km away at the Gila National Forest in New Mexico.

Does it matter? Yes. Here's the answer in two parts. The stars have been part of our heritage for millennia; they have given us the calendar, timekeeping and navigation. (OK — these days we depend on computers rather than stars.) And the heavens are as much a part of our landscape as are lakes and mountains on Earth. You wouldn't build a city in an area of such natural beauty. So why destroy our view of the heavens, which are an inspiration to us all?

Satellite's eye view of the world's hottest light pollution spots

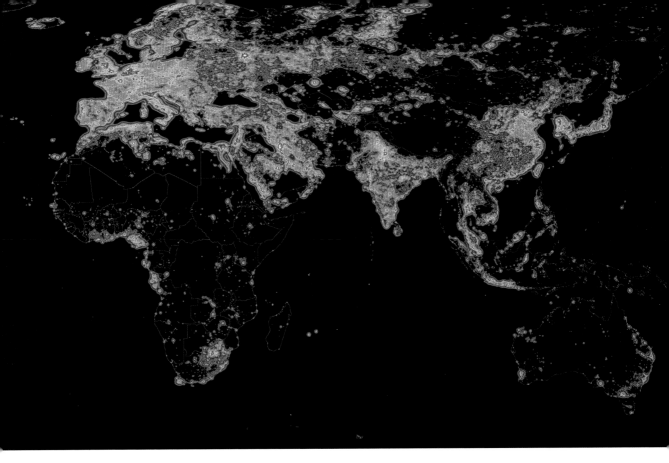

Just as important, light pollution is a waste of money and energy — and it's a major contributor to global warming. In the U.S. alone, the highly respected International Dark Sky Association (IDA) has calculated that 30 percent of all outdoor lighting is wasted, mostly as a result of unshielded lights. The bill? $3.3 billion a year — and the release of 21 million tons of carbon dioxide (from coal-fired power stations) to increase the greenhouse effect. To compensate for the electricity that simply lights up the sky, you'd have to plant 875 million trees annually to mop up the extra CO_2!

There are many ways to combat light pollution, both financially and aesthetically. Shield streetlights so that light goes down, not up.

Use low-pressure sodium as a light source for streetlights — it's far less polluting and far more efficient. (Unfortunately, the new energy-efficient LED streetlights don't necessarily help, as they produce blue light that's easily scattered into the atmosphere then back to us as light pollution.) Plus — cut industrial pollution, which is a menace when it comes to scattering stray light.

And the aesthetics? Head to a Dark Sky Park. These are places where light pollution is virtually nonexistent. Carefully monitored by the IDA and other organizations, these sites are a joy to behold, with stars stretching down to the horizon. Check out "Dark Sky Preserve" on Wikipedia, and . . . enjoy!

Is the North Star the brightest star?

The short answer is no. The North Star's claim to fame is that it always lies due north in the sky, and so it helps us navigate. That's because it lies right over the Earth's North Pole; but it's only a matter of luck that there's a star in that direction at all.

The Southern Hemisphere isn't really blessed with a pole star. Sigma Octantis, barely visible to the unaided eye, hovers some distance from pole-prime position. But it does have the honor of being the faintest star featured on any country's flag (Brazil).

In fact, the North Star (also known as the Pole Star and Polaris) is a medium-bright star, about the same brilliance as the stars making up the familiar Big Dipper (or Plough). In the image to the left, the North Star is at the top — the shortest bright arc — and, as you can see, it really does not stand out.

This gorgeous "star trails" photo is a long-exposure image taken with a camera that was fixed, pointing northward with the shutter left open for several hours. As the Earth rotated, the stars wheeled around the north pole of the sky, tracing out arcs of light. Because it's closest to the pole of the sky, the North Star draws the shortest trail. As you can see, though, it's not exactly at the pole. If it were, it would be just a point of light.

The North Star will be closest to true north on March 24, 2100. That's because the Earth's axis is wobbling like a spinning top in space. This phenomenon — called precession — takes 26,000 years to complete, and we get to see different pole stars during the cycle. Hang on for 12,000 years or so, and we'll be blessed with a dazzling pole star in the shape of the brilliant star Vega.

The North Star (top right), circled by its neighbor stars

Why do stars twinkle? Do planets twinkle too?

There's nothing more romantic than spending an evening together under the gently twinkling stars. But hang on, there's a bright "star" that's different: it's shining with a steady glow like a tiny lantern. It's a planet. Even with your naked eye, you can tell a planet from a bright star because it does not twinkle.

The stars twinkle as the shifting air currents bubble over them. Planets aren't affected in the same way and shine with a steady light.

Here's why . . . Living below Earth's churning atmosphere is like lying at the bottom of a swimming pool and looking upward. The water swirls overhead, creating pockets of good and bad visibility. Small objects floating on the surface, like leaves, are smeared out the most, flickering in and out of sight. Larger items, such as beach balls, aren't affected nearly as much by the turbulence of the water, and we see them easily.

And so it is with the heavens. The stars — so far away that they appear as no more than points of light — twinkle as the shifting air currents bubble over them. Planets, which appear as tiny discs, aren't affected in the same way, and shine with a steady light. But both planets and stars twinkle when they're close to the horizon and we see them through a greater thickness of roiling atmosphere.

Though the twinkling of the stars — technically called scintillation — may be romantic, it gives astronomers a serious headache, as it blurs out fine detail in astronomical images. One solution is to create an artificial "star" above the telescope using a powerful laser that lights up atoms in the upper atmosphere. Astronomers tweak the telescope's focus to keep the fake star crystal clear, and this automatically sharpens up the real stars in the same field of view.

A brilliant laser beam at Spain's Teide Observatory shoots into the upper atmosphere to help sharpen the telescope's view of the twinkling stars.

Can I see satellites?

You bet! When you go out on a clear night, you'll only have to wait a quarter of an hour or so to spot your first celestial interloper. Seasoned spotters can see around 100 satellites per night.

Almost all the satellites you'll see will be in Low Earth Orbit (LEO). There are around 500 satellites in LEO today. It's a useful orbit for Earth observation, spying and communications. It's also great for satellite spotters — the satellites are low enough and appear large enough to reflect sunlight.

How to locate your quarry? Look for an object, traveling at around the speed of a plane, moving silently and steadily against the starry backdrop. It might not be very bright (compare it to the stars in the Dipper), but if it has solar panels, it may suddenly flare as they catch the sunlight.

Satellites have a disconcerting habit of disappearing. Don't panic. They haven't fallen from the sky, but have moved into the Earth's shadow!

By far the most spectacular satellite to spot is the International Space Station. Some 400 km up, it takes a leisurely 4–5 minutes to cross the sky. With its huge bulk and enormous solar panels, it can appear as bright as the planet Venus. Do what we do: go down to your local pub or bar, get everyone outside and spot the space station! We all get very excited, and give a symbolic wave to the astronauts as they fly over.

No local country bar? Never fear. Search for "Spot the Station" on the Internet, enter the details of your location, and NASA will do the rest.

Trail of the International Space Station captured on camera

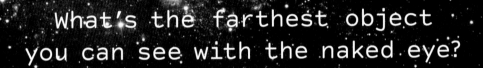

What's the farthest object you can see with the naked eye?

It all depends on your eyesight — and the degree of light pollution you suffer (see page 18).

From downtown in a city, forget it — you'll just see a handful of the brightest stars.

Go to the suburbs, and you might spot the Orion Nebula. It's a glowing star-factory 1,300 light years away from Earth, and a fabulous sight through a small telescope (see Chapter 9 for more on nebulae).

Country-dwellers have a wealth of distant objects to gaze upon: nebulae, star clusters and even galaxies well beyond our local star city. Without a doubt, the farthest here is the Andromeda Galaxy (see Chapter 11) — 2.5 million light years away.

Then it's up to your eyesight. Many observational astronomers have excellent eyesight, but they're divided on this one. The Triangulum Galaxy, M33, should *just* be visible to the unaided eye. But the small spiral galaxy is very low in brightness: some amateur astronomers can see it, while others can't.

The answer? Get yourself into a desert, and M33 is all yours. You've taken the record — the Triangulum Galaxy is three million light years distant!

The Triangulum Galaxy, M33 — the most distant object visible to the unaided eye

When gazing from a dark place, why do I see some parts of the sky glowing?

Nigel was driving from Grand Canyon to Flagstaff, Arizona, one spring evening after sunset, keeping an eye on the twilight glow in his rearview mirror so that he could stop for a bit of stargazing when the sky was really dark.

But it just didn't seem to fade. He stopped and looked behind. It wasn't light from the sunset at all, but a faint glow that stretched in a tall pyramid up through the constellations of the zodiac.

The zodiacal light — a ghostly radiance that even the faintest moonlight can obliterate — extends along the plane that planets travel in the Solar System. You'll see it best in spring and autumn, when the zodiac is at its steepest angle to the horizon. It's caused by tiny particles of dust between the planets that scatter light from the Sun. Most likely, the dust is generated by a family of ageing comets held in the gravitational thrall of mighty Jupiter (see Chapter 8 for more on comets).

One leading expert on these microscopic cosmic rocks is a colossal rock star — of the musical kind. Brian May — guitarist with Queen — started a PhD about the zodiacal light, but abandoned it when his musical career went supernova. He eventually completed his doctorate at Imperial College, London, 37 years after he had commenced it! "A Survey of Radial Velocities in the Zodiacal Dust Cloud" resulted in his becoming "Dr. May" in 2008. He attributed its success to the zodiacal light becoming "trendier" — a result of our incredible spacecraft forays into the Solar System.

A second much brighter river of light you'll see crossing the night sky is the Milky Way. It's as far from the Solar System as you can imagine. We live in a lens-shaped, spiral galaxy (see Chapter 11), and while its closer stars are scattered evenly over the sky, perspective makes its more distant inhabitants converge into a band. Using his tiny "optik tube," Galileo discovered this in 1610 and declared the Milky Way to be "a congeries of stars."

You can even see the Milky Way from the suburbs of a city. Sweep its length with binoculars, and you'll understand what Galileo meant.

Wide-angle shot of the Milky Way (left) and the zodiacal light (right) above the Very Large Telescope in Chile

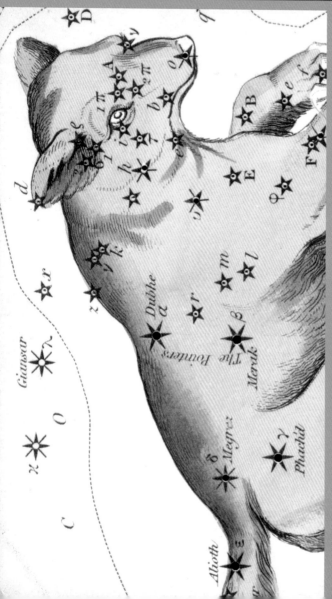

MYTHS AND LEGENDS

2

Was Stonehenge an observatory?

In the hours before dawn on Midsummer's Day, we waited expectantly among the dark pillars of Stonehenge. As the sky grew brighter, a procession of chanting figures clad in pure white robes paraded to the heart of the ancient stones, staring outward to the distant Heel Stone. Their song grew to a frenzy at the moment the Sun rose above the sacred megalith.

We were lucky enough to be at the world's premier stone circle, witnessing the most famous sunrise of the year. But the builders of Stonehenge didn't leave a manual, so can we be sure that it was actually built to celebrate the Sun?

In the 1960s, Gerald Hawkins — a British astronomer working in Boston, Massachusetts — decided to solve the riddle of Stonehenge once and for all by using the newfangled electronic computer. (To put his machine in context, your cell phone has thousands of times the computing power of his IBM 7090.)

Hawkins programmed the computer to work out the direction of lines joining up all the stones, and found that an amazing number pointed to spots on the horizon where the Sun and Moon rose and set on significant dates. He concluded that Stonehenge was a sophisticated astronomical observatory used for predicting eclipses, the most terrifying of all sky sights.

Sad to say, scientists later found that Hawkins had programmed his computer wrong. And the result, as the IT guys say, was "garbage in, garbage out."

Only the line linking the center of Stonehenge with the Heel Stone and the Midsummer sunrise stands up to scrutiny. But a line points in two opposite directions: if you look the other way, you'll

Midwinter sunset at Stonehenge, in Wiltshire, England

see the Sun setting at Midwinter. Which is right?

Recently, archaeologists have excavated the "wild town next to Stonehenge where the builders partied," as one newspaper headline put it. They've found the remains of countless ancient feasts where people from all over Britain came and celebrated once a year. Among the debris are bones and teeth from young pigs — mostly slaughtered at the age of nine months. Because wild pigs are born in springtime, these remains prove that the great celebrations at Stonehenge actually took place in the depths of winter.

So the modern Druids visiting Stonehenge at Midsummer's dawn were mistaken. Stonehenge was built so that people could stand at the outlying Heel Stone and watch the Sun *set* behind the great edifice (as you can see in the picture) on the afternoon of *Midwinter's* Day.

How did ancient people think the universe was created?

All cultures have asked themselves the fundamental questions: "Where did we come from?" and "How did it all begin?" In the absence of the hard evidence we have today, our ancestors came up with some colorful myths.

The Egyptian civilization — which lasted two millennia — had an elaborate mythological structure revolving around their many gods and goddesses. First was Atum, who came into existence simply by calling his own name! After vomiting up his brother and sister, Atum's dynasty began to grow, resulting in the god of the Earth (Geb), and the sky goddess (Nut) arching over him. All the people of Egypt were descended from the children of Nut and Geb.

The Chinese believed that the universe was created in a giant cosmic egg, containing *yin-yang* — which comprised everything and its opposite: male-female, dark-light, heat-cold.

The Aztecs of Mexico, renowned for their gory legends, tell the story of their gods Quetzalcoatl and Tezcatlipoca. They, rather ungallantly, pulled the goddess Coatlicue down from the heavens and tore her in two, creating the sky and the Earth. Understandably, she was unhappy at her treatment, and demanded frequent sacrifices of human hearts.

According to Judeo-Christian texts, "In the beginning, God created the Heaven and the Earth." Working hard for six days, he established day and night; seas, land and plants; the Sun and Moon; and sea creatures and birds. On day six, he made animals and the pinnacle of his creation: human beings. He set aside the seventh day for a well-earned rest, making a Sabbath holy to Jews and Christians.

The Egyptian goddess Nut arches her body over the Earth to create the heavens.

When people believed in a flat Earth, what supported it?

Back in the days when early people didn't travel much beyond their own country, it was natural to think the world was flat. But why didn't it just fall down through space?

Ancient Indian pundits came up with a solution. The Earth was carried on the back of four powerful elephants. But — hang on — why didn't the elephants fall through space? They must be standing on something large and totally immovable: the World Turtle.

Once the idea traveled to the sceptical scientists of Europe, however, the turtle was in trouble. In Victorian times, an eminent astronomer gave a public lecture in which he described the machinations of the Solar System, with the Moon orbiting the Earth and the planets circling the Sun. At the end, a lady in the audience put up her hand and said, "I have a much simpler theory," and described the elephants and the World Turtle. The scientist countered, "So, what does the turtle stand on?" And his adversary decisively stated, "It's turtles all the way down!"

In his chart-topping book series *Discworld*, British science fantasy author Terry Pratchett had a better solution. His imaginary world — as its name suggests — is round and flat, and supported by four patient elephants standing on a giant turtle. Pratchett's turtle has a more elegant way of staying buoyant: it endlessly swims through space.

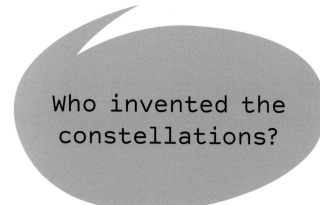

Who invented the constellations?

Look out on a January night, and you'll see the figure of a great hunter — Orion — sketched out by the stars like a figure in a connect-the-dots puzzle book. Nearby is a bull and Orion's two faithful dogs. Elsewhere in the sky, you'll find a great bear, two water snakes and a giraffe — not to mention a unicorn and even a chemist's furnace!

As the variety of names suggest, there wasn't one single constellation-maker. Many different peoples have sketched their ideas on the sky over history. The great water snake Hydra (opposite, top) was created around 2800 BC, and the crow (Corvus) and the cup (Crater) were in place by ancient Greek times. The sextant, air pump, owl and cat were added in the centuries following 1687 — though in the 20th century the astronomical authorities axed the bird and the cat.

The Great Bear (opposite, bottom) may be the oldest star pattern, dating back to the Stone Age, because native North Americans — who left the Old World over 13,000 years ago — see the same bear-shape as the Europeans.

The Elamites, a mysterious civilization living on the shores of the Persian Gulf around 3000 BC, used images of bulls and lions to personify the power of their rulers. They created the signs of the zodiac, the constellations marking the Sun's path in the sky.

About the same time, the Minoans of Crete drew up gigantic sprawling constellations, such as Hercules, to help them navigate the Mediterranean.

The Greeks created celestial tableaux from their myths. A line of three stars depicted the princess Andromeda chained to a rock, with a W-shape of five stars as her mother, Queen Cassiopeia. Nearby you can spot the outline of the sea-monster that's about to devour her, and the hero Perseus flying to her rescue.

In the 17th century, cartographers making maps of recently discovered regions of the Earth also depicted new patterns in the sky, largely from the southern stars never visible from Europe. The Dutch mapmaker Petrus Plancius created a celestial peacock, goldfish and chameleon. He also amputated the legs of the ancient constellation of the Centaur to form the distinctive Southern Cross. The French astronomer Nicolas-Louis de Lacaille brought the skies into the age of science with a pendulum clock and — fittingly — a telescope.

Over 4000 years of inspiration, projected onto the sky. Above: Hydra surrounded by smaller constellations. Below: Ursa Major — the great bear.

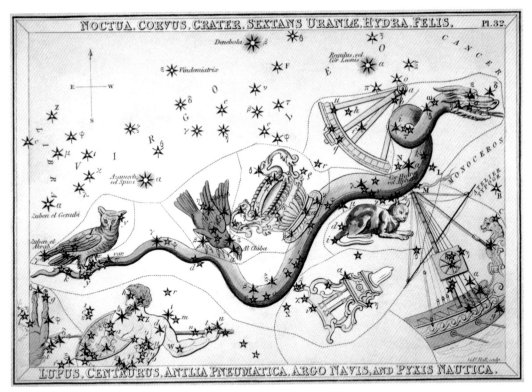

NOCTUA, CORVUS, CRATER, SEXTANS URANIÆ, HYDRA, FELIS. Pl.32.

LUPUS, CENTAURUS, ANTLIA PNEUMATICA, ARGO NAVIS, and PYXIS NAUTICA.

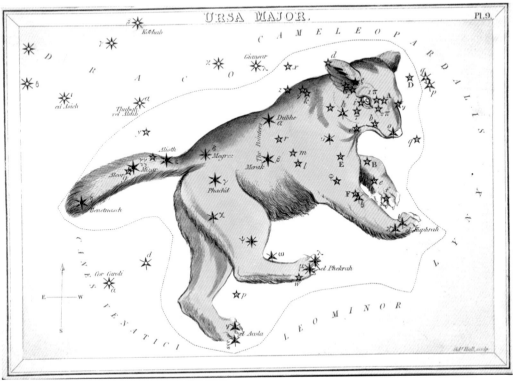

URSA MAJOR. Pl.9.

Look at a Chinese constellation map, and you'd be forgiven for getting lost. Their heavens are packed with a myriad of tiny star patterns, unrecognizable from the sprawling constellations that we've become used to in the West. Orion and the Big Dipper are the only ones you'd pick out.

Understandably, the Chinese interpretation of the constellations is very different from ours. Take the familiar W-shaped constellation of Cassiopeia: to us, a queen sitting in a chair. Not to the Chinese — they saw this small star-pattern as three constellations! First and foremost was a charioteer being pulled by a team of four horses. Then there was a path through the mountains (the glowing band of the Milky Way running through Cassiopeia). If that wasn't enough, they also threw in an *alternative* route through the mountains!

What do you do when you're faced with a sky with too many stars? This was the problem facing the Aboriginal Australians in their pitch-dark deserts. The answer? Join up the *dark* patches in the heavens to create a very different kind of sky lore. Their favorite constellation (one depicted on several stone engravings) was the emu, made up

of dark rifts along the band of the Milky Way (see Chapter 11). Its head was the Coalsack — a dark cloud near the Southern Cross — while its body meandered down to the stars of Scorpius.

The Aboriginal people weren't averse to "connecting the dots" as well. They saw Orion, which they named Julpan, as a canoe being rowed by three fishermen (Orion's Belt to us). One of them caught a forbidden fish (the Orion Nebula). All three were cast into the heavens by Walu, the Sun-woman, for their misdemeanor.

In North America, the indigenous peoples have a rich sky lore. Each tribe has its own legends; the following one comes from the Chinook tribe. Like the Aboriginal Australians, they saw Orion as a canoe — actually, two canoes. The big canoe was represented by Orion's Belt; the small one was his dagger.

The canoes were competing in a challenge to catch a salmon from the Big River (the Milky Way). The small canoe was winning the race and eventually spotted their quarry in the middle of the river. It was Sirius — the brightest star in the sky.

Who named the stars? Can I have a star named after me?

Naming the stars was largely done by Arab astronomers during the so-called Dark Ages in Europe (8th–12th centuries). Take Algol, in the constellation Perseus. In legend, Perseus was the superhero who slaughtered Medusa, the Gorgon whose gaze could turn any onlooker to stone. He used her head to great effect when rescuing Andromeda from Cetus, the ravaging sea-monster. Possibly because the astronomers noticed something odd about Algol — it dims regularly every three days — they named it "the head of the ogre," *al-ghūl*.

The name of Betelgeuse in Orion has suffered many mistranslations from the original Arabic. One came out as "The Armpit of the Central One"! This, alas, has been toned down to "The Hand of Orion."

A rare Chinese survival is Tsih, the central star in Cassiopeia (which the Chinese saw as a charioteer): it means "The Whip."

Greek and Latin names also grace the firmament. Spica, in Virgo, is Latin for "ear of corn," while Arcturus — just below the Great Bear — comes from the Greek "guardian of the bear." Sirius, the brightest star in the sky, also has a Greek name: it means "glowing" or "scorcher." The Greeks knew that the first appearance of

Sirius in their skies heralded the hottest days of summer. And this wasn't good news: plants wilted, men weakened and women became aroused. Sirius was regarded as a malignant influence, emitting "emanations." Anyone affected by these was said to be "star-struck."

As for having a star named after you — no way! The only authority that can name stars is the International Astronomical Union (IAU). It will *not* consider requests to name stars from members of the public.

Most stars aren't named at all; being below naked-eye brightness, they're just allocated a catalog number.

A ghoulish constellation: Perseus holds Medusa's head, marked by the star Algol .

What was the Star of Bethlehem?

Well, we're pretty sure of what it *wasn't*. The Wise Men didn't observe a supernova — a brilliant exploding star of the kind depicted on Christmas cards today. Ancient Chinese astronomers were observing the sky assiduously every night, and they recorded nothing amazing at the time.

And it wasn't Halley's comet — another popular choice for the star — because that came around in 12 BC, several years too early. (Scholars dispute the exact date when Jesus was born: it was most likely around 6 or 7 BC.)

In those days, the phrase "wise men" meant astrologers, busy studying the planets as they moved around the zodiac. And in the year 7 BC, Jupiter, king of the planets, and Saturn — symbolising the Biblical region of Jesus' birth — met up three times in the constellation of Pisces, the sign of the Jewish people according to astrologers. A remarkable celestial portent that the King of Jews (as Jesus was referred to in the Bible) was on his way? Not so fast: conjunctions of Jupiter and Saturn aren't particularly rare.

Most worryingly, there's no evidence for the Star of Bethlehem apart from a few brief mentions by St. Matthew. Bible scholars tell us that Matthew had one overriding mission: to convince fellow Jews that Jesus wasn't just a prophet but was the anticipated Messiah. His advent called for supernatural events to rival the birth of the great patriarch Abraham, when "one great star came from the east and ran athwart the Heavens."

So, perhaps the Star of Bethlehem never actually existed. As a devoted biographer of the new messiah, Matthew may simply have felt compelled to write a brilliant star into his script!

It wasn't Halley's comet, as depicted by Giotto in 1304.

Is there any truth in astrology?

No. So why is it so popular? First, it's very lucrative for astrologers. It's also a solace to people who have trouble finding direction in life. After all, the astrologers pump out their platitudes with such pseudoscientific conviction that the readers feel there *has* to be something in it.

To be fair, serious astrologers distance themselves from the horoscopes churned out by the newspapers. They pride themselves on star charts drawn up according to the time of a person's birth, in which they plot the positions of the Sun, Moon and planets against the stars making up the zodiac (the constellations that follow the plane of our Solar System).

That's a mistake. Western constellations were drawn up thousands of years ago, and since then, Earth's axis has wobbled around, changing where the Sun, Moon and planets lie among the zodiacal constellations (see Chapter 1).

Then they use impressive words to describe the apparent line-ups of the planets against the constellations, and with each other. Did you know you may have a quincunx in your birth chart? Wow! It's when two planets (or the Sun or Moon) lie 150 degrees apart — and this can have huge effects on your health, love life, travel plans, work, etc.

All of this mumbo jumbo is backed up by curious symbols with absolutely no scientific relevance. Push astrologers to explain the *mechanism* behind their predictions, and they can't.

Astrology is a great way to make money and provide people with a dubious decision-making system, but it's trash. In the 1950s, psychologist Michel Gauquelin undertook a huge survey of birth charts. He found that there was absolutely no relation between people's achievements in life and the position of the Sun, Moon or planets.

He also offered newspaper readers a free horoscope. Ninety percent declared that it was completely accurate. In fact, Gauquelin had sent them all the same personality profile – based on the birth chart of a mass murderer!

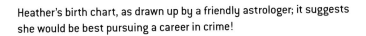

Heather's birth chart, as drawn up by a friendly astrologer; it suggests she would be best pursuing a career in crime!

Do all cultural traditions see a face in the Full Moon?

The Full Moon is a big friendly face in the sky, his eyes and mouth clearly marked by large dark expanses against his glowing white cheeks, chin and forehead. In the Western world, we're so used to seeing the Moon's markings this way that it comes as quite a shock to learn it's not the only interpretation.

In medieval Europe, the dark markings depicted the shape of Cain — the first murderer, according to Biblical tradition — carrying a bunch of twigs on his back. Scandinavians have seen two children carrying a water jug. Incan myths relate the story of a young woman who sprang up and into the Moon, while several myths scattered throughout the Americas speak of a woman bent with age.

But around the world, the most common shape is the hare in the Moon (see image). According to a Buddhist tradition, he's offering to be eaten alive to save a starving priest. In China, the hare is pounding rice in a mortar. For the Khoikhoi in South Africa, he's bringing a message of hope down to Earth, but he mangles his sentences. The Moon punishes him by splitting his lip, while the hare retaliates by scratching the Moon with the marks we see to this day.

The Hare in the Moon. We've emphasized the dark areas to show the shape more clearly.

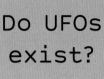

Do UFOs exist?

Yes, they certainly do! And we've seen some ourselves . . .

Now, you may be surprised that we — as astronomers — say that. But the meaning of UFO is simply "unidentified flying object." Once, we spied a distant silvery disc after sunset and had no idea what it was. This was a UFO — until we focused our binoculars on it and saw that the "disc" was a distant flock of flying geese. (Then it became an IFO — you can work out what that means!)

Of course, many people have reported UFOs, and claimed that these "flying saucers" are spacecraft piloted by aliens. When they're investigated, most turn out to be IFOs — most commonly, sightings of weather balloons, aircraft or the planet Venus. Others turn out to be deliberate hoaxes.

In a court of law, the defendant has to be found guilty "beyond reasonable doubt." In all the cases we've seen, no one has proven *beyond reasonable doubt* that any UFOs are piloted by living beings.

So, there may be plenty of unidentified objects in the sky, but none of these UFOs are alien spacecraft.

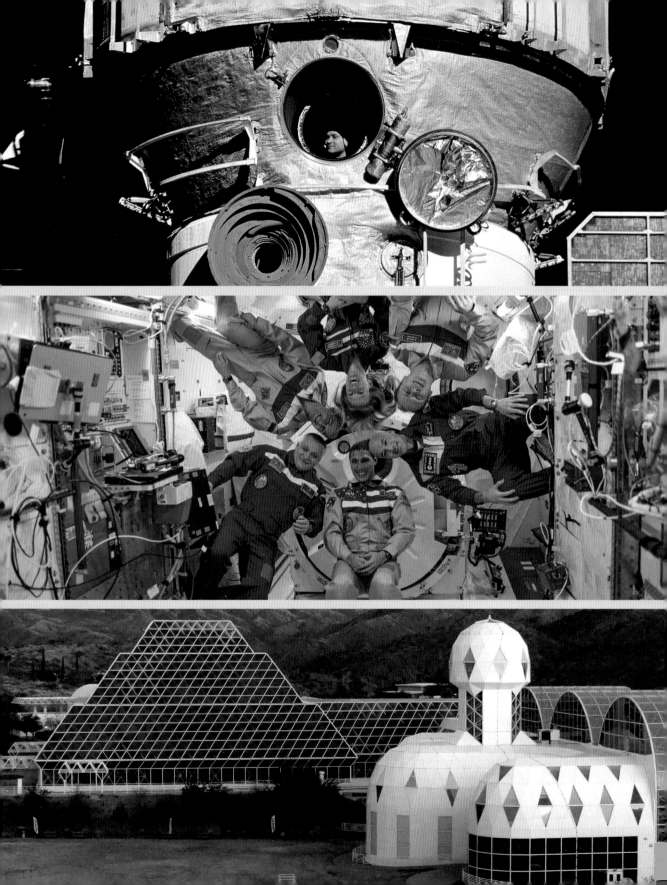

SPACE TRAVEL 3

What's the most powerful rocket ever built?

NASA's giant new rocket, the Space Launch System (SLS), is powerful enough to blast a crew of 22 full-grown elephants into space — and, in case you're worried about animal welfare, they have the volume of nine school buses to play around in. (That's according to NASA's own statistics.) In reality, it's designed to fling a human crew way beyond the reach of Earth's gravity, eventually to Mars.

NASA claims that the SLS is the most powerful rocket ever built — but is it?

Those 22 elephants come to a total of 130 tons. That's a lot more than the lift capacity of the famous space shuttles, which could loft a mere three elephants (24 tons). And the most powerful current rocket, SpaceX's Falcon Heavy, can loft 64 tons.

But way back in the 1960s, NASA claimed its Saturn V Moon rocket could launch 140 tons, which seems more powerful than the SLS. In fact, NASA has changed what it means: when talking about the Saturn V, they included the mass of the upper stage and its fuel. Its actual payload was only 122 tons (a couple of elephants less than the SLS). So the SLS does deserve the laurels in terms of getting stuff up into space.

But its performance relies on the power of several individual rockets: a core stage and strap-on boosters which fire together, and then an upper stage. If you're interested in the pure power of a single rocket, then every American first stage (including the SLS) is surpassed by the massive muscle of the Moon rocket that the Soviet Union built and test-fired four times at the height of the Space Race: the mighty N-1 with its 30 engine nozzles. (It's a shame that all the test flights failed, though, and the N-1 never made it into space . . .)

NASA's Space Launch System is powerful enough to blast a crew of 22 full-grown elephants into space.

The new Space Launch System is NASA's mightiest rocket.

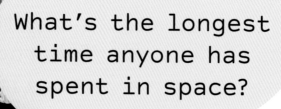

What's the longest time anyone has spent in space?

Many astronauts and cosmonauts have taken several trips to orbit, and the spacefarer who's amassed the most time "out there" is cosmonaut Gennady Padalka. Since 1998, Padalka has been on five separate missions, to Mir and the International Space Station, and has totalled 879 days in space — almost two-and-a-half years! As we write this, Padalka is scheduled to fly to the ISS again late in 2018, to add another six months to his already out-of-this-world record and bring his total up to over 1,000 days in space.

Only two Americans feature in the top 20 spacefarers when it comes to total time in space. One is Scott Kelly, who hit the headlines in March 2016 when he returned to Earth after almost a year on the ISS with fellow "One-Year Mission" cosmonaut Mikhail Kornienko. They were guinea pigs for American and Russian doctors who want to probe how the human body responds to a really long dose of zero gravity like astronauts will experience on a trip to Mars.

And Kelly brought an extra bonus for the medics — his twin brother. Mark Kelly stayed on the ground while Scott was whizzing overhead, so NASA could compare directly how gravity versus weightlessness affects two identical bodies.

O n January 8, 1994, cosmonaut Valeriy Polyakov blasted off on a mission to the Russian space station Mir. He arrived back on Earth on March 22, 1995, after more than 14 months in space — to be precise, a record 437 days and 18 hours. Our picture shows Polyakov after a year in orbit, watching America's space shuttle *Discovery* arrive for a welcome visit.

Cosmonaut Valeriy V. Polyakov watches *Discovery*'s rendezvous with Mir.

What happened to the Russian space dogs?

The Russians used dogs for their first space forays — chosen from strays wandering the streets of Moscow. They were selected for their resilience, their tolerance of hot and cold conditions and their desire to lie down as much as possible!

Little Laika was such a stray. She was launched on *Sputnik 2* on November 3, 1957 — who can forget images of her plaintive little face looking out of the spacecraft's window?

The Russians had no way to recover her, and knew she would die. The technicians sadly kissed her nose before strapping her in for launch. Alas: within hours she was dead — a result of over-heating and stress.

The next Russian pooches in space were much luckier. Strelka and Belka were launched on *Sputnik 5* on August 19, 1960. On getting back safely to Earth, Strelka proceeded to have six puppies, one of whom, Pushinka (Fluffy), was presented by Nikita Khrushchev to John F. Kennedy.

Pushinka herself had four puppies (called "pupniks" by Kennedy), and her descendants are still alive and thriving across the U.S. today.

A descendant of space dog Strelka was presented as a gift to U.S. president John F. Kennedy.

Belka and Strelka commemorated on a postage stamp

How many astronauts have been killed?

The sad total is 22 astronauts.

Early 1967 saw the first American and Russian fatalities taking place within three months of each other. On January 27, astronauts Gus Grissom, Ed White and Roger Chaffee entered the Command/Service Module of *Apollo 1* at the Kennedy Space Center, Florida, in a routine ground-test. But a spark from an electrical cord ignited the nylon in the astronauts' suits in the pure oxygen atmosphere. Controllers heard a cry of "Fire!" followed by a scream. By the time they could open the capsule, they found the astronauts dead.

In the Soviet Union, cosmonaut Vladamir Komarov flew the first mission of the new *Soyuz* craft. It had been plagued by problems — and everything that could go wrong, did. On April 24, 1967, Komarov reentered Earth's atmosphere successfully. But the main parachute wouldn't deploy, and the brave cosmonaut was killed.

Fast-forward to *Soyuz 11* on June 30, 1971. Three cosmonauts — Georgy Dobrovolsky, Vladislav Volkov and Viktor Patsayev — had finished a successful mission visiting the world's first space station, Salyut 1. On reentry, after a record three weeks in space, a breathing ventilation valve was jolted open. The cabin depressurized in seconds. The recovery team found all three astronauts motionless in their seats — dead from asphyxiation.

The next tragedy took place on January 28, 1986. Christa McAuliffe, the first teacher in space, boarded the U.S. space shuttle *Challenger* with her fellow crew members: Dick Scobee, Michael Smith, Judy Resnik, Ellison Onizuka, Ronald McNair and Gregory Jarvis.

Exactly 73 seconds after lift-off, there was a huge explosion, and *Challenger* disintegrated in a grotesque fireball. We know that some of the astronauts *did* survive the explosion, because emergency equipment had been activated. But they perished instantly in the 200 g impact with the Atlantic Ocean.

Challenger had launched in freezing conditions, at −2°C (28°F). The rubber O-rings joining segments of the solid rocket boosters strapped to the shuttle had lost their flexibility at this low temperature, and searing-hot gas had escaped through a joint. This blowtorch burned through the supports and ignited the flammable mix of hydrogen and oxygen in *Challenger*'s huge external tank.

On February 1, 2003, tragedy struck a Shuttle again. It was the 28th mission of *Columbia*, with

a crew of Rick Husband, William McCool, Michael Anderson, Ilan Ramon, Kalpana Chawla, David Brown and Laurel Clark.

Unknown to the crew, at launch a suitcase-sized piece of foam insulation from the external tank had hit the Shuttle's left wing, punching a 15 × 25 cm hole. As *Columbia* reentered Earth's atmosphere, hot gases penetrated the hole, broke up the wing and set the Shuttle violently out of control. It disintegrated in a trail of burning debris as thousands of horrified people watched. They were witnessing the death of the first space shuttle to orbit the Earth — and of its seven valiant crew members.

Private spaceflight suffered its first fatality on October 31, 2014, as pilots Peter Siebold and Michael Alsbury were testing Virgin Galactic's *SpaceShipTwo* over California's Mojave Desert. Alsbury mistakenly unlocked a lever to "feather" the spacecraft's tail, tipping it upward to its reentry position. But with the rocket engine still firing, the premature feathering broke *SpaceShipTwo* in half. Alsbury died in the accident; Siebold parachuted safely to Earth.

But let's not end on a gloomy note. Over 550 people have now traveled to space, so the fatality rate for astronauts is under 4 percent.

Haunting image of space shuttle *Challenger*'s explosion, January 28, 1986

How dangerous is space junk?

Our 60-year legacy of space exploration has led to some spectacular triumphs. But success has its downside: in this case, the ever-increasing swarm of space debris that litters Earth orbit.

This isn't just a cosmetic issue. "Space junk" is hazardous: traveling at tens of thousands of kilometers an hour, it can destroy satellites and threaten the safety of occupied spacecraft, such as the International Space Station.

In 2016, the U.S. Strategic Command tracked nearly 20,000 artificial objects in orbit. And these were just the fragments big enough to show on the radar; it's estimated that there are over 170 million objects smaller than a centimeter circling our planet.

Where does it come from? The largest culprits are discarded upper stages of rockets, which ferry satellites into space. Then there are the satellites themselves. Ageing rockets and satellites can explode as their components deteriorate, mixing their volatile fuels. And satellites can collide — there were five collisions in 2016 — which adds to the rising tally of space pollution.

Then there are weirdo candidates. A space glove, pliers, a toolbox, a spatula, and even flecks of paint orbit the Earth. In 2016, a window on the International Space Station was struck by a fleck thousandths of a millimeter across. It gouged a hole 7 mm in diameter. "I'm glad the window was quadruple-glazed!" observed astronaut Tim Peake.

Most space debris is destroyed when it eventually spirals downward and burns up in the atmosphere. But the biggest junk can survive all the way down to Earth's surface. In 1969, for instance, bits of a Russian spacecraft injured five sailors on a Japanese ship. The most massive piece of space machinery to strike our planet

A decaying rocket explodes, showering Earth orbit with debris.

was Skylab, America's first space station. It crashed into an area near Perth, Australia, in 1979. The spacecraft, which weighed over 77,000 kg, was a victim of a sudden glitch of solar storms. As a result, our upper atmosphere expanded, increasing the drag — and the huge space station hurtled to Earth.

Space junk faces a number of challenges from Earth. The latest was Kounotori (Stork), a ten-year collaboration between the Japanese Space Agency, JAXA, and the fishing net manufacturer, Nitto Seimo. Launched in 2016, a tether based on fishing net plaiting technology was designed to ensnare space junk in an electromagnetic field. The idea? Slow down the junk and let it burn up in the Earth's lower atmosphere. Alas: Kounotori failed in 2017.

But there are plenty more ingenious solutions on the drawing board. Our favorite — proposed (somewhat tongue-in-cheek) at a space debris conference — was to send a huge satellite made of cosmic chewing gum into orbit!

Over half of all astronauts suffer from space sickness. Like motion sickness on the Earth, it's caused by disorientation. When you're weightless, the balance organs in your ears aren't registering the same orientation as your eyes are showing you: some astronauts feel sick when they see fellow crew members walking upside down, or suddenly catch a glimpse of the Earth at a strange angle through a window.

Rusty Schweikart, on *Apollo 9*, was one of the first astronauts to admit he was spacesick: "suddenly I had to barf. . . . And, I mean, that's not a good feeling." And it didn't give the other two crew members in the cramped capsule a good feeling either! Schweikart went on to test the new lunar spacesuit: fortunately he didn't throw up then.

And vomiting in a spacesuit is every astronaut's nightmare. If your barf floats in front of your face, you could breathe it in and suffocate in a most gruesome way. It's never actually happened, though one astronaut — whom NASA refuses to name — was slightly sick when donning a spacesuit. Space sickness usually abates after two or three days in space. That's why astronauts never undertake a spacewalk immediately after they arrive at the International Space Station.

I t was April 1985, and Senator Jake Garn was unwell. Unfortunately for him — and his crewmates — Garn wasn't at home in bed, but was whizzing around the world in space shuttle *Discovery*. The first "space tourist," Garn's role was largely that of a medical guinea pig. And he was providing a bonanza for the doctors, suffering nausea, cold sweating, fatigue and loss of appetite, along with a pallid skin and a raised temperature.

This was the worst recorded case of space sickness. And, informally, NASA astronauts now rate their experiences with the "Garn scale": one-tenth of a Garn is bearable, but no one ever wants to suffer space sickness at the one-Garn level!

"Houston, I've got a problem . . . "

What do astronauts eat?

It was the sandwich that almost grounded NASA's *Gemini* piloted spaceflights. During the *Gemini 3* mission in March 1965, John Young produced a corned beef sandwich he'd smuggled on board. Gus Grissom took a mouthful, before realizing that crumbs were floating around in weightlessness — and could have caused a short-circuit in the spacecraft's electronics.

Looking back, it's hard to blame the astronauts for being tempted by tasty real food. NASA was so worried about food or drink floating around out of control that the astronauts originally had to live on mashed food they squeezed from a tube, like toothpaste.

Today, space food has gone haute cuisine. Delivered to the International Space Station in cans, in plastic pouches or freeze-dried, it's reconstituted on board with hot water. Russian cosmonauts have a choice of over 300 dishes, ranging from goulash and borscht to jellied pike perch.

NASA also provides a wide choice of options, including breakfast sausage links, chicken fajitas, barbecued beef brisket, macaroni and cheese, and cherry-blueberry cobbler. For condiments, you have mustard, ketchup and mayonnaise. Salt and pepper — which could pose a danger free-floating — are only available in a liquid form.

More suited to veggie tastes (like ours!) is the meal prepared for Italian astronaut Luca Parmitano: eggplant parmigiana with mushroom and pesto risotto and caponata, followed by tiramisu for dessert.

And tortillas are big in the space kitchen. You can serve food on them without worrying about cleaning up, and — unlike bread — tortillas do not produce crumbs.

When he visited the ISS in 2015, British astronaut Tim Peake wasn't going to be deprived of his traditional bacon sandwich. Leading chef Heston Blumenthal worked for months to develop a space "bacon sarnie" that would pass all the regulations. Soon after he arrived, Peake enjoyed a successful space sandwich made with sticky brown bread — and, of course, a cup of tea.

Astronaut Ellison S. Onizuka eats a gourmet meal on the Earth-orbiting space shuttle *Discovery*.

Has anyone had sex in space?

R umours abound, of course. But no one's telling. However, fish and fruit flies have been getting up close and personal to each other for decades.

NASA insists that no one has ever had sex on any of their spacecraft. But in 1994, the space shuttle *Columbia* carried an experiment to investigate whether fish could reproduce in weightlessness. The tiny Medaka fish (Japanese rice fish) successfully mated, conceived and gave birth in space. All the offspring were perfectly normal.

Now the Japanese have an aquarium on the International Space Station so that scientists can observe how zero gravity affects fish and their spacefaring descendants. They hope that it will help us understand medical conditions on Earth — like muscular atrophy — that happen quickly in space.

No one bats an eyelid about fish having sex in space. But *geckos* — oh my god! On July 18, 2014, the Russians launched a microgravity satellite, Foton-M, with a crew of five geckos and a colony of fruit flies. Its express purpose? To study mating in weightlessness.

The world's media went bonkers. "Space Sex Mission" and "Russian Space Sex Geckos," screamed the headlines. One newspaper mused about scientists' cameras watching "live lizard porn." The geckos became international celebrities on TV and the Internet. The five space lizards — four females and one male (more than one, and you're asking for trouble) — seemed happy at first. They even played, which they seldom do on Earth.

But it was not to be a happily-ever-after story. On its fourth orbit, the satellite ceased responding. Although communications resumed, Foton-M wasn't behaving according to plan. Controllers brought it down early and discovered — to their horror — that the geckos were dead.

What went wrong? It seems that the geckos froze in their box after the communications problem. So: no little geckos. At least their putative parents had their moment of fame.

As for the fruit flies, in a different compartment on the satellite, it was mission successful. In the words of CNN: "Now — lucky us — the world has more fruit flies."

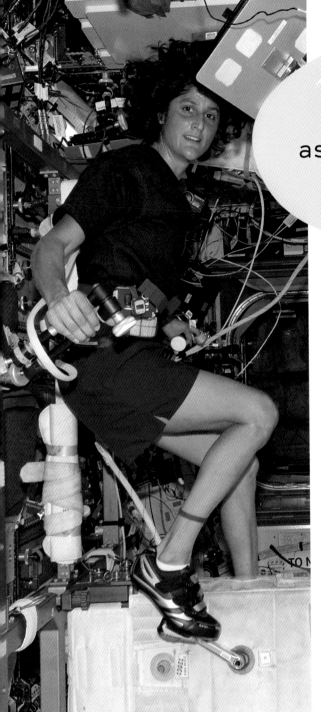

How much do astronauts have to exercise?

If you don't like the gym, then a long space trip isn't for you. Mission controllers say that after eating and sleeping, exercise is an astronaut's most important activity — spacefarers on the International Space Station must spend up to two-and-a-half hours working out every day!

In weightless conditions, your muscles gradually atrophy; if you don't keep them toned, you won't be able to stand when you return to Earth. In addition, your bones gradually lose calcium, becoming weak and brittle. Exercise helps prevent this calcium loss.

The Russian gym on the International Space Station features an exercise bike and a treadmill with a strap to hold the cosmonaut onto the track. The American crew have their own versions of these, along with a complex machine called the Advanced Resistive Exercise Device — with pistons and flywheels that simulate weightlifting on Earth — and the Miniature Exercise Device, a prototype of the compact fitness equipment needed on the smaller spacecraft that will take astronauts on the long mission to Mars.

In 2016, Tim Peake ran a marathon on board the International Space Station, clocking a time of 3 hours 35 minutes on the treadmill.

Work out or fade out . . . Flight engineer Sunita Williams exercises on the Cycle Ergometer with Vibration Isolation System on the International Space Station.

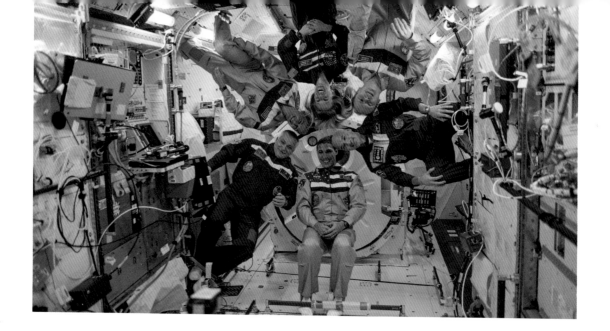

Does weightlessness mean there's no gravity in space?

It's a bit confusing. We casually say astronauts are "weightless" on the International Space Station as they float around freely. It seems that gravity doesn't exist there.

But hold on. The ISS is in the thrall of Earth's gravity: that's why it orbits our planet rather than flying off into space. In fact, the gravity of our planet reaches out as far as the Moon — keeping it in orbit — and beyond.

So what's going on? Imagine standing outside the ISS and watching it zoom past, with an astronaut on a spacewalk outside. This astronaut is orbiting the Earth just like the space station, so she flies alongside the ISS without being pulled toward or away from it. Her colleague inside the ISS is also traveling in orbit along with the space station around him, so he doesn't move toward or away from the floor or walls either — he floats around inside, without being drawn to the floor.

Here's another way of looking at it. Nothing is holding the ISS up, so Earth's gravity is pulling it downward. The only reason it doesn't crash on our heads is that it's traveling sideways: the space station is traveling so fast that its curved path downward exactly matches the round shape of the Earth underneath.

The astronauts aboard are falling along with the ISS. When an elevator or a theme park ride drops suddenly, you momentarily feel weightless. In the falling ISS, that weightlessness persists. That's why scientists often refer to the apparent weightlessness of orbit as "free fall."

Which way is up? The astronauts on the ISS float in weightless formation.

How do astronauts go to the toilet?

However you look at it, this could end up being pretty messy. In space, everything is weightless, and you really don't want the stuff that comes out of your body just floating around out of control!

During the Apollo missions to the Moon, the astronauts (all male) would pee into a flexible tube that went straight out into space. For "number two," they had a plastic bag. It was held to their buttocks by glue, which proved painful for the hairier astronauts . . .

The space shuttle had a proper sit-down toilet, with moveable thigh grips so the astronaut wouldn't float away in the middle of a session. Again a hose was used for urine, with a different shaped cup at the front for male and female anatomy. Solid waste was pulled down into the toilet bowl by suction: a fan flung it around the inside of the container, where it could stick and dry out. But sometimes this material would break up and float back into the cabin as a cloud of disgusting "fecal dust."

Aboard the International Space Station, current astronauts have a sit-down toilet, with a hose for collecting pee. The urine is recycled into drinking water. Astronauts joke that "Yesterday's coffee becomes today's coffee!"

The astronauts' "number two" is collected in a plastic bag (not glued to them this time). There's a selection of toilet paper — the rough Russian variety and softer tissues for Americans and Europeans — which ends up in the bag too. After use, the astronaut seals the bag, and suction pulls it into the base of the toilet. When the container is full, it's loaded — along with other waste — into an empty space freighter, which is dispatched back into the atmosphere to burn up: the space-poop ends up as a glorious meteor!

Bathroom visits weren't a pleasure on the Mir space station.

How big is the International Space Station?

The short answer is: HUGE! At 109 m long, the International Space Station is bigger than a football field, and its 16 solar panels have an area of eight basketball courts. It's also the most expensive object ever constructed, at an estimated cost of $100 billion.

Traveling at 27,500 km/h — 10 times faster than a rifle bullet — the ISS orbits the Earth in just 90 minutes. In one day, it covers the distance to the Moon and back. It took 155 space flights to launch all the components of the ISS, and 201 spacewalks to assemble and maintain them. The whole thing has a mass of 420 tons, which is roughly the same as 320 cars.

As we write this (August 2017), 222 people have lived on the ISS, from 18 different countries. The astronauts enjoy more living space than you'd find in a six-bedroom house: they have two bathrooms, a gym and a 360-degree panoramic window for out-of-this-world views of planet Earth.

Watch out — 420 tonnes of metal coming this way at 27,500 km/hour!

Is the Great Wall of China visible from space?

Back in 1932, *Ripley's Believe It or Not!* ran a page about China. Along with amazing facts such as "All the cash in China has holes in it" and "The word Peace is indicated by a pig under a roof," we find a description of the Great Wall of China as "The mightiest work of man — the only one that would be visible to the human eye from the Moon!"

This claim was actually checked out within the lifetime of their youngest readers. After *Apollo 11* visited the Moon in 1969, Neil Armstrong reported back on the Great Wall of China: "It is not visible from lunar distance." Buzz Aldrin added, "You have a hard time even seeing continents."

But surely it's easy to spot the Great Wall from space, if you're talking about astronauts in low orbit around the Earth? Crew members on the International Space Station can certainly eyeball many contenders for "the mightiest work of man," including cities, giant opencast mines, a huge greenhouse complex in Spain and the wakes left by large ships.

The Great Wall is elusive, though: it may be long, but it's not very high or wide, and you're best off looking for the shadow it casts when the Sun is low. A few astronauts claim they've glimpsed the Wall, but there's always the chance of mistaken identity. On the Skylab 4 mission, William Pogue thought he saw the Great Wall with his unaided eye, but binoculars showed him it was actually the Grand Canal near Beijing.

Let's leave the last word to Yang Liwei, China's first *taikonaut* (Chinese for astronaut), who orbited the Earth in 2003: "The Earth looked very beautiful from space, but I did not see our Great Wall."

Easy to see from here, but maybe not from space.

What would happen to an astronaut exposed to space?

Don't hold your breath! That's the number one tip to follow (otherwise your lungs may explode).

But first, what are the dangers of space to the naked human body? It's seriously cold out there, but don't worry about freezing: the vacuum around you does not conduct heat (that's how a Thermos works too). There's also a lot of dangerous radiation from the Sun, but you'll be dead before you suffer lethal sunburn or radiation sickness.

Despite the rumours, your blood won't boil inside your veins. It's true that water boils at a lower temperature when the pressure is lower, but even in space your blood is held in by your arteries and veins. If your blood pressure is a normal 120/80, the boiling point of water is higher than 45°C (113°F), and your body temperature of 37°C (98.6°F) isn't hot enough to make it boil.

What will kill you is the vacuum. Without any air pressure around you, the gas in your lungs is exerting a huge force outwards, like an inflated balloon. If you hold your breath, your lungs will explode, rupturing your innards and forcing air into your blood vessels, which will stop your heart and brain from functioning. It's called "explosive decompression."

So — let the air out. The gases in your blood escape rapidly into the vacuum now in your lungs, depriving your brain of oxygen. Within 12 seconds, you black out. Your muscles alternately spasm and become paralysed; your heart begins to pound, and then your pulse rate falls and your circulation slows down to a crawl. Once your heart stops, you are a goner.

The good news? If someone can get you into a recompression chamber within 90 seconds of your exposure to space, you're likely to make a full recovery.

His spacesuit was all that protected astronaut Bruce McCandless from a horrible fate.

How dangerous is a human trip to Mars?

The answer is: VERY. It's not a three-day hop to the Moon. It's a six-month journey across the hazardous wastes of space, eighteen months on an unpredictable world, and a further six months to return to Earth safely.

There's no shortage of volunteers — scientists and astronauts alike are vying for their place on the first mission to Mars. But it will be the biggest challenge that humankind has ever attempted.

The chief risk is radiation. On the journey to Mars, space weather will be a major hazard. The Sun, our normally gentle local star, can be a fierce beast — hurling out energetically charged particles when it peaks in its eleven-year cycle of activity.

And more distant stars will also affect our future explorers. Exploding stars — supernovae — shoot deadly cosmic rays into space, and these heavy particles have lethal consequences. So be prepared for some dead or severely injured astronauts. As a NASA spokesperson observed: "We fully expect to get some wet noodles back."

Protecting the spacecraft with a water shield will help the radiation hazard; but, when added to the craft's fuel, it will add to its weight. The solution will be to send a fully fueled spacecraft that will wait for them on Mars, ready for their return.

Then there's the prospect of spending 18 months on Mars, until the Earth and the Red Planet get into alignment for the shortest journey back. How will we explore such an alien world? How will we cope with its unpredictable dust storms?

Now we're faced with the biggest question of all: *Who* will go to Mars? Everyone agrees that it will need to be an international crew; no one nation could afford the price tag. And it will need to be of mixed gender. An ideal crew would

Biosphere 2: Arizona's practice-run for living on Mars

consist of six, possibly three married couples — then you can avoid the old issue of sex in space.

One crew member, at least, would need to be a doctor. You can't afford untreated appendicitis on a trip to Mars.

But in the end, it all comes down to psychological strains. "Imagine traveling around England in a motor home with five other people for three years," a psychologist explained to us. "Without ever going outside."

The crew for the first human mission to Mars will have to be very carefully evaluated — this is the key to its success. Experts look to the example of Antarctic explorer Ernest Shackleton, who chose his teams for their sense of humor in situations of extreme stress.

When will it happen? Most people would agree with Barack Obama, in a speech he made in 2010. "By the mid-2030s, I believe we can send humans to orbit Mars and return them safely to Earth. A landing will follow. And I expect to be around to see it."

Who will be the first to land on Mars?

We're certainly not taking bets on this one! According to their own hype, a commercial company will land a human on Mars ahead of any government. Their secret? A one-way ticket — so there's no need for the cost and complexity of a return mission. The Dutch organization Mars One has already picked its astronaut team, and plans to create its Mars colony in 2031. The only problem is that they have yet to raise the money . . .

The founder of Tesla, Elon Musk, has both deep pockets and a working rocket. His company, SpaceX, has been sending cargo to the International Space Station, and is beginning to launch astronauts. He's designing a powerful booster — the BFR (Big Falcon Rocket) — which would each carry 100 people to the Red Planet to create an eventual colony of a million settlers. Though most experts believe he's wildly optimistic, Musk reckons he could be dispatching ships to the Red Planet by 2030 . . .

NASA is looking at a mission that will take astronauts to Mars and return them to Earth. It has both the rocket power and the crew capsule — in the shape of the new SLS and Orion capsule — to wing a human mission on the long trip to the Red Planet by the 2030s. But the estimated cost of an *Apollo*-type assault on Mars ranges from $100 billion to a scary $1 trillion.

And successive administrations have blown hot and cold. In 1989, President George H. W. Bush announced "a journey to another planet — a manned mission to Mars." Roll on 15 years, and President George W. Bush described the next step in space exploration as "human missions to Mars." Another 12 years later, in 2016, President Barack Obama proclaimed the next chapter of America's story in space as "sending humans to Mars by the 2030s."

In an interplanetary version of the hare-and-tortoise race, while NASA stumbles about at the whim of its political masters, China is slowly and steadily progressing in space. Already, they have their own space station, and have launched a program to land a crew on the Moon. No one doubts that they will then head to Mars. Many experts predict that the first national flag to fly on the Red Planet will be China's five gold stars on a red background.

NASA's 1964 vision of a Mars landing

Can people live on Mars?

You can't just step out of your space capsule and take a deep breath to celebrate your arrival. Mars' air is so thin — less than 1 percent of Earth's atmospheric pressure — that you'd immediately die from decompression, just like an astronaut exposed to space (see page 61). In addition, it's made of unbreathable carbon dioxide gas, the average temperature is –55°C (–67°F) and, with no ozone layer, you'd be fried by the Sun's ultraviolet rays.

But the good news is that it's easy to make a Mars base. Cover the modules with soil to protect the astronauts from radiation, and use solar energy to heat them. The same energy can be used to split the ice in Mars' surface layers into breathable oxygen (at a pressure we can breathe) plus hydrogen to use as fuel.

And you can grow crops in a Martian greenhouse. Dutch researchers have raised peas, tomatoes, radishes, and rye (for making bread) in simulated Martian soil, while the International Potato Center has identified a variety of spud that would survive there — so you could indeed live on potatoes, like the stranded astronaut in the movie *The Martian*.

It would be a total waste of resources to keep animals, so the Martian colonists will be treated to a tasty veggie diet.

Further in the future, some enthusiasts talk about terraforming Mars — turning it into an Earthlike world. First, pump huge amounts of greenhouse gases into Mars' atmosphere (the opposite of what we are trying to achieve on Earth!) to raise the temperature. That will heat Mars' currently frigid poles and release huge deposits of frozen carbon dioxide. This gas will add to the greenhouse effect and thicken the atmosphere to a pressure similar to that found on Earth's high mountains. Cover Mars with alpine plants to convert some of this gas to oxygen, which will form a protective ozone layer. Equipped with simple oxygen masks, like mountaineers, our descendants on Mars will be able to roam freely over the Red Planet.

Artist's impression of buried modules and future greenhouses on the Red Planet

What's the most distant man-made object?

As we write these words, a NASA website is telling us that their unpiloted *Voyager 1* spacecraft is exactly 20,858,257,012 km from home. That means the hardy space probe (launched in 1977) is 140 times the Earth's distance from the Sun — or 140 astronomical units (AU), for short. That's four-and-a-half times farther out than the most remote planet, Neptune, and three-and-a-half times more distant than Pluto.

Next in line is *Pioneer 10* (which is no longer in touch with Earth) at 119 AU, followed by *Voyager 2* at 115 AU — like its record-setter sister ship, still sending back valuable scientific data. Even traveling at the speed of light, radio messages from the two *Voyagers* take the best part of a day to reach Earth!

But even that is puny on the scale of interstellar distances.

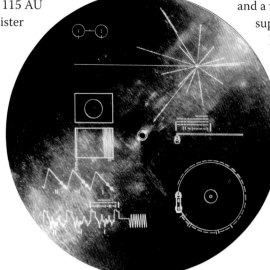

The light from the nearest star spends over four *years* on its journey.

Just in case they get picked up by curious aliens, each *Voyager* carries a greeting from Earth in the form of the most sophisticated data storage available in the 1970s — an LP record! It looks like a vinyl disc, but is made of gold-coated copper that won't degrade in space. A clever alien could run the enclosed stylus around the groove and listen to tracks ranging from Bach to Chuck Berry. Other sections code images of the Earth, including snowflakes and a woman shopping at a supermarket. There's also an hour-long recording of the brainwaves of one of the team, Ann Druyan, as she meditated on the human condition, from violence and poverty to falling in love (she later married the leader of the project, Carl Sagan).

Top of the interstellar charts: the cover of *Voyager*'s LP record carries a user's manual.

Can you buy a ticket to space?

Yes, and there's a choice of operators. First in the game was the U.S. company Space Adventures, which sells tickets to fly with a Russian *Soyuz* crew to the International Space Station. So far, seven individuals have taken a one- to two-week vacation in orbit, including a Microsoft Office developer Charles Simonyi, who's been up twice. The price is a commercial secret, but it's similar to the amount NASA pays the Russians for a *Soyuz* seat — a cool $81.7 million.

If your sights are set farther out, Elon Musk's company SpaceX is offering a trip around the Moon (but without a landing), on his Falcon Heavy — the rocket that launched his Tesla roadster to Mars. He hasn't revealed the price, but it won't be a cheap ticket!

If you have only $250,000 in your vacation fund, then turn to Virgin Galactic. Founded by British entrepreneur Richard Branson, it's building a fleet of small spaceplanes that will fly six passengers at a time on a suborbital hop — shooting above the 100-km threshold for a few minutes in space. Named *SpaceShipTwo*, the winged vehicle is based on the first private spacecraft, *SpaceShipOne*, which flew above 100 km three times in 2004.

Jeff Bezos, founder of Amazon.com, has also stepped up to the plate. His Blue Origin company has built a reusable rocket that will loft six people over a height of 100 km in a capsule that then parachutes down to Earth. It's the same profile that America's first astronaut, Al Shepard, flew in 1962, and Bezos has called his launcher *New Shepard.* He intends to start selling tickets to space in 2018 — price, so far, undisclosed.

Virgin Galactic's *SpaceShipTwo* is carried aloft by a twin-hulled plane at Spaceport America, New Mexico.

Can spaceships travel faster than light?

Chemical rockets can't get us much faster than *New Horizons*, but a plasma rocket — which uses electric power to eject million-degree gases — can build up to higher speeds, in theory up to 15 percent of light speed over interstellar distances.

Even better, don't take any fuel with you. Instead, use powerful lasers at home to push a lightweight space probe equipped with a "sail" out into space. That's the idea behind the Breakthrough Starshot, which plans to send a miniature craft to the nearest star at 20 percent the speed of light.

But, can any rocket break through the natural speed limit of the universe? After all, Albert Einstein himself proved nothing can travel faster than light.

There are plenty of ideas around for some kind of a warp drive, as used in the *Star Trek* television show — just try trawling the Web. But few stand up to scientific scrutiny. The most promising has been devised by Mexican physicist Miguel Alcubierre. Inspired by *Star Trek*, he proposes that a spaceship could create a "bubble" in the space-time around it. According to relativity theory, the bubble can travel faster than light — and Alcubierre believes the spaceship could surf the bubble to break the light-speed barrier.

N ASA's *New Horizons* holds the record for fastest man-made object: when launched in 2006, it sped into space at 58,536 km/h — 20 times faster than a rifle bullet. While the *Apollo* astronauts took three days to reach the Moon, *New Horizons* crossed the Moon's orbit after only nine hours!

But that's nowhere near the speed of light (moonlight takes just over a second to reach us).

New Horizons speeds past Pluto.

What will happen to us when the Sun dies?

We are all doomed! No, that's not a quote from the dismal guy standing on the street corner. It's a serious warning from hardheaded astronomers.

One day, the Sun will die. And then life on Earth can't survive. Even though the final act is five billion years in the future, it's never too early to start planning!

The Sun is gradually growing hotter as it ages, and in around three billion years its heat will be intense enough to evaporate Earth's oceans. The steamy atmosphere will trap more of the Sun's energy, and — combined with carbon dioxide released from rocks — a runaway greenhouse effect will push the Earth's thermostat to boiling point. Time to get away...

And our next home will be easy to reach. Mars will warm up naturally, with lakes and an atmosphere thick enough to protect us. Just add plenty of grass and trees to turn its carbon dioxide air to oxygen, and we're well ensconced for a billion years.

But the solar furnace will keep ratcheting up. Eventually Mars will go the way of the parched Earth. We'll have to search for a new home, farther out. The outer planets aren't for humans: they are all gaseous, with no solid surface. Their biggest moons are a good bet. But Jupiter's large companions lie within its deadly radiation belts.

Saturn's great moon, Titan, beckons. It already has a thick atmosphere, composed mainly of nitrogen — like the Earth's air. NASA's *Cassini* spacecraft has probed beneath its clouds (see image above) and found lakes of methane on an icy surface. Four billion years from now, Titan will be warming up — the methane will be natural gas, while the surface will be awash with welcome seas.

But nothing can stop the final act, when the dying Sun puffs off its outer layers to leave a tiny fading core — a white dwarf. We will have to leave Titan and the Solar System behind, and voyage out to planets orbiting nearby stars that are still in the prime of life.

Saturn's moon Titan: Desirable residence, circa 5 billion AD?

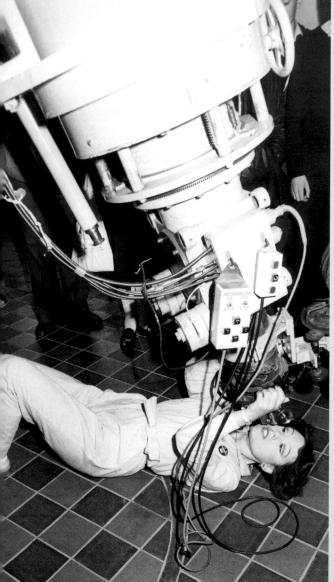

TELESCOPES 4

What's the best telescope for observing from my backyard?

It's a jungle out there! If you go to a star party, it seems as if telescopes come in every possible shape, size and price.

But let us offer a few signposts through the undergrowth. First, beware of buying a telescope from a general catalog or gift website: their telescopes are cheap for a reason — they aren't very good!

And your interest in the sky can affect what kind of telescope you choose. If you want to fly low over the Moon and magnify the planets to view them in intimate detail, you'll be best off with a refractor. This is the traditional design, with a big lens at one end and an eyepiece at the other. Make sure you buy a telescope that's achromatic, which means it has a double front lens to supress false color fringes. Even better, but more expensive, is an apochromatic refractor.

On the other hand, if "faint fuzzies" — like nebulae and galaxies — are your style, then seek out a reflector. This has a mirror at the bottom to collect and focus light, and — because mirrors are cheaper than lenses — you catch more precious light for your buck. The biggest telescopes you'll find at a star party are Dobsonians, which are reflectors stripped of all the trimmings so they are just huge "light-buckets."

Combining the best of both worlds are catadioptric telescopes. Here a mirror does the hard work of collecting the light, but a lens at the top end of the tube helps to focus a sharp image. The result is a very powerful but compact 'scope. These Schmidt-Cassegrain and Maksutov telescopes are often computer-controlled, which takes out the hard work of finding your astronomical quarry. Just key in "Saturn" or "Crab Nebula" on the GOTO keypad, and the telescope will point itself straight at a delectable sky sight.

A menagerie of telescopes at a star party: refractors (left) and a catadioptric telescope (right)

Who invented the telescope?

One of the best-known "facts" in the history of astronomy is that Galileo invented the telescope. Only he didn't (see the next question for what he actually did).

The patent on the first telescope was taken out in 1608, a year before Galileo began stargazing. In the Dutch city of Middelburg, an optician called Hans Lipperhey (confusingly, he sometimes signed his name Lippershey) described his new invention as "a certain art with which one can see all things very far away as if there were nearby, by means of sights of glasses."

Lipperhey had held one spectacle lens in front of another, and found that he could magnify a distant view. It's something that must have happened before, but with the crude lenses of the time, the combination usually led to a large but fuzzy image. Lipperhey added one crucial component: a circular ring that stopped light from passing through the outer part of the front lens, which caused most of the distortions.

We don't know if Lipperhey himself looked at the sky, but a curious Dutch citizen must have peeked upward, because a contemporary pamphlet states, "even the stars which ordinarily are invisible to our sight and our eyes, because of their smallness and the weakness of our eyesight, can be seen by means of this instrument."

The news spread rapidly, and "Dutch trunkes" were soon being sold at fairs all across Europe. One came into the hands of Thomas Harriot, an earlier English explorer of the New World who had introduced potatoes into Europe. Harriot drew the first sketches of the Moon's rough surface as revealed by the telescope. And his friend Sir William Lower came up with our favorite description of the lunar terrain: "she appears like a tart that my cooke made me last weeke; here a vaine of bright stuffe, and there of darke, and so confusedlie all over."

Spying out the view at Middelburg with the newly invented "Dutch trunke"

What did Galileo do?

Fiery-haired Italians had something in common, or so it seems. Galileo Galilei (1564–1642) and Antonio Vivaldi (1678–1741) knew no boundaries: the latter in music, the former in science.

Galileo was a brilliant physicist and mathematician. Although he trained as a medic (under the insistence of his equally determined composer father, Vincenzio), his real love was for experimental science.

> **Galileo was a brilliant physicist and mathematician. Although he trained as a medic his real love was for experimental science.**

Always abrasive, he was annoyed at the dogma laid down by the ancient Greeks. They taught that heavier objects fall faster than lighter ones. Galileo allegedly proved them wrong by dropping two balls of different weights off the Tower of Pisa. They hit the ground together, leading to the foundations of Isaac Newton's theory of gravity.

Teaching math at the prestigious University of Padua, two things hit him: Copernicus' ideas of a Sun-centered universe, and the invention of the telescope in Holland.

Now he got into a nice little business. Using exquisite glass from the Venetian island of Murano, he made a telescope. He managed to convince the warring authorities in Venice that it was an excellent enemy-deterrent, being able to spot invading ships long before they were visible to the unaided eye. The doge (ruler) agreed, and doubled his salary.

But ever the scientist, Galileo wondered about the astronomical applications of his "optik tube." He pointed it at the Moon, and discovered that it was pockmarked. So — the heavenly bodies *weren't* perfect. Looking at Venus, he noted that it showed phases — just as it should if it orbited the Sun. And Jupiter was circled by tiny dots — moons in their own right, controlled by the giant planet. And the Sun was spotty . . .

All this convinced Galileo that Copernicus had been right: we are merely a planet orbiting the Sun.

But his discoveries were starting to make waves in the Vatican. The papal authorities, who believed in the supremacy of the Earth, were

moderately relaxed that Copernicus' heretical ideas were kept within the confines of the scientific community. But — horror! Galileo was publishing his findings in the people's language of *Italian*, not in learned Latin.

The pope had had enough. Galileo was flung into lifelong house arrest in his villa in Arcetri, near Florence. Even the belligerent scientist was careful not to practice much astronomy later in life.

But word of Galileo's findings leaked out. After his death, the great scientist became a cult figure. News of his work blazed across Europe, triggering astronomical research — and the rest is history.

Galileo displays his astronomical telescope to Leonardo Donato, Doge of Venice, as later depicted by French painter Henry-Julien Detouche.

What's a planisphere?

What's the best way to get started in astronomy? The answer is a planisphere (also known as a plane sphere or star-dial) — use it for a year, and the stars will be yours. These flat star-dials are remarkably cheap. They have a window, showing the stars and constellations visible to the unaided eye, and a dial around the edge.

Using one is simplicity itself. First, select the month and date. Then line up the time (inner circle) with the date. Hold the planisphere above your head (the pivot at the center marks the position of north or south in the sky), and presto — that is what's on view in the heavens for that exact date and time.

There are two things to remember about planispheres. First, get one for your correct latitude — otherwise the stars won't be in the right place. Second, you won't see planets marked; they move across the sky over the course of a year. But this is a fantastic tool to introduce you to the heavens. Now you're ready for the next step.

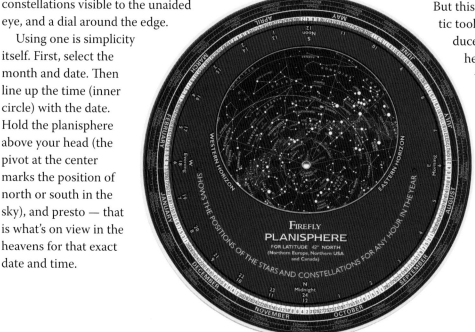

A planisphere is the ideal way for learning the stars by dialing the sky.

Are binoculars any use for astronomy?

After a planisphere, binoculars are your next move. You now know the sky, but try this experiment. Look at the stars through a toilet-paper tube. See how it restricts your view of the heavens? If you borrow a telescope from a pal, you'll find that it limits your view even more. The answer? Binoculars. Yes, they do restrict your view, but at least you can look away and see roughly where they were pointing.

Most amateur astronomers, even if they own a massive telescope, carry binoculars in the car all the time. Stands to reason: you can whip them out if you're in a clear, dark spot, or simply if you want panoramic vistas of the sky, which a myopic telescope — with its close-up views — won't give you.

Good, sturdy binoculars will give you fantastic sights of the Milky Way, double stars, nebulae, the moons of Jupiter, star clusters and the nearest galaxies. Support them on a fence to minimize wobbling from your arms. If they're really big, you'll need a robust mounting.

What to choose? Ignore ads that promise huge magnification. They might look slick, but these tiny-lensed binoculars will simply deliver an enlarged blur.

Binocular sizes come in a code that gives the magnification and the aperture of the lenses. 7 × 50 binoculars have a magnification of 7 and lenses that are 50 mm across. These, or 10 × 50s, are ideal for a beginner. And they won't break the bank!

Binoculars are great — both for stargazing and birdwatching.

Can an amateur astronomer still make useful discoveries?

Unlike professional astronomers, who have limited access to telescope time (or satellite time) in observatories, amateurs are free to roam the heavens whenever they want. And that's how they make serendipitous discoveries.

On February 23, 1987, in New Zealand, retired miller Albert Jones was out with his telescope monitoring variable stars — stars that change in brightness. It's painstaking work, but it adds to our knowledge of how stars become unstable late in life.

The sky was clouding over, but Jones decided to take a peek at one of his favorite objects before putting away his tools. When he looked at the Large Magellanic Cloud — one of the Milky Way's satellite galaxies — he was in for a shock. There was a brilliant star that hadn't been there before.

Soon clouded out, Jones picked up the phone to alert other observers. Word spread quickly around the world: Albert Jones had discovered a supernova, an exploding star (see Chapter 9). It was the first to be visible to the unaided eye for nearly 400 years. In recognition of his outstanding contributions to astronomy, the Queen awarded the modest Jones an OBE (Officer of the Order of the British Empire).

Backyard astronomers are adept at finding supernovae. Holding the world record is Scotsman Tom Boles, who has discovered an astonishing 149 "stellar suicides" (as of 2016) in distant galaxies.

Backyard astronomers are adept at finding super- novae. They also excel at discovering comets

Amateur astronomers also excel at discovering comets (see Chapter 8). The brilliant comet that graced our skies in 1997 was found independently in the U.S. by experienced comet-hunter Alan Hale and rookie astronomer Tom Bopp. Hale — a professional astronomer who looks for comets for fun — made his first discovery at last. "When I'm not looking for one, I get one dumped on my lap!"

Bopp, stargazing in the desert with fellow enthusiasts, stumbled over the fuzzy patch when scanning nebulae in Sagittarius. The group checked; it wasn't on their star charts. Bopp then

had the daunting task of calling the International Astronomical Union to report that he'd found a comet. They called back. "Congratulations, Tom. I believe you discovered a new comet."

Hale-Bopp reigned supreme until 2007, when Scottish astronomer Rob McNaught — veteran of more than 50 comet discoveries — found his latest comet at Australia's Siding Spring Observatory. It was a sensation in the Southern Hemisphere. Brighter than Venus, and sporting a fan-shaped tail, astronomers described it as "like looking at ten comets with tails all at once."

The message is clear: whether you're an experienced or a beginning stargazer, you never know when you might make a discovery. And if you find a comet, there's an added bonus: it gets named after you — so you achieve cosmic immortality!

Supernova 1987a (right, near center), next to the awesome Tarantula Nebula

Do carrots help you see in the dark?

If you're suffering a severe vitamin deficiency, the answer is probably yes — because your sight is impaired and carrots will boost your vitamin A.

The human eye needs a good supply of vitamin A to create a substance called visual purple (rhodopsin) that's essential for seeing clearly in low light levels. But anyone with a normal diet has plenty of vitamin A, and eating more carrots won't make the eye any more sensitive.

This urban myth was deliberately created by the British Air Ministry during the Second World War. They were determined to keep secret the fact that they were tracking German planes at night with their newly invented radar system, so they claimed that British pilots boosted their night vision by eating carrots!

Perhaps, if you're a rabbit . . .

Can I make an astronomical discovery through the Internet?

I t used to be called armchair astronomy: stargazing for people who prefer to keep warm with a book on matters celestial, rather than freeze in the back garden.

With the Internet, all that has changed. Now you can stay warm *and* do astronomy. The name of the game is Citizen Science.

Professional astronomers acquire much more data than they can possibly analyze. So they share it with keen amateurs by sending their results to your computer.

You can keep on reading your book if you want, and let the computer do the work. Simply download the program, and it will run in the background without any intervention. You can search for alien radio signals (see Chapter 14), investigate the shapes of asteroids or plot stars in the Milky Way.

But it's much more fun to be hands-on: you'll get guidance from the professionals and act as a member of their team. You'll be seeing images from space probes and telescopes that no one has seen before. You'll be checking out planets orbiting other stars, the distribution of craters on the Moon, or clouds on Mars.

A very popular site is Galaxy Zoo (www. galaxyzoo.org), where you can help the pros classify the shapes of unusual galaxies. That's where Dutch schoolteacher Hanny van Arkel was when she discovered Hanny's Object. It's a very unusual gas cloud ripped from a galaxy (see image above).

Astronomers have now found other small shining clouds in intergalactic space — named *Voorwerpjes* after the Dutch for "small objects" — but they're very rare. The finger points to explosive quasar activity (see Chapter 11) in the main galaxy that has long since died away. Hanny's Object is now making stars — look at the top of the cloud where it's being fed gas by its parent galaxy.

So you don't have to brave the cold to make a cosmic discovery. And with Citizen Science, you can make real progress at astronomy's cutting edge.

Hanny's Object (green) was discovered as part of the Galaxy Zoo project.

How can I become an astronomer?

Anyone can become an astronomer. You just need a fascination with the heavens and a passion for stargazing. It's not like brain surgery, where there's no way you can be an amateur. But follow our tips on planispheres, binoculars and telescopes, and you're part of a huge community who loves the sky.

You'll find your own particular fascination in astronomy. The Moon, the planets, comets, meteors, variable stars, deep sky objects — the list is endless.

But what if you want to go pro? This a different life. It means a university degree, with a high enough grade to do research at college or a major organization like NASA. You need to be pretty brilliant at math and physics if you don't want to go belly-up at the equations of hydrostatic equilibrium (those that power the stars) in your second year at university.

Believe us: it's worth it. It opens up a whole new world of astrophysics: cosmology, dark matter, black holes, supernovae, exploding galaxies and the Big Bang.

But if the lure of astronomy is its pure beauty, stay with your stargazer status. Amateur and professional astronomers respect each other, and often work together. And the beautiful 19th-century telescope that I'm showing off below? Well, it's used by amateurs today. So come on in — the water's lovely!

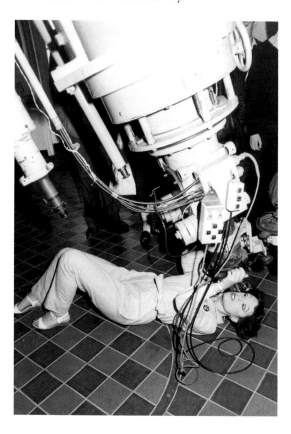

Heather demonstrating the mighty 28-inch refractor at the Royal Observatory, Greenwich.

Why do astronomers put telescopes on mountaintops?

Twinkle, twinkle little star; how I wonder what you are . . . " The gentle twinkling of the stars may be romantic, but astronomers — trying to find out what the stars are — actively hate it!

As we mentioned in Chapter 1, stars twinkle because their light travels down through layers of the Earth's shifting atmosphere to reach the ground. And the churning air blurs out details that you hope to see through a powerful telescope.

It may seem obvious that you'd get a better view from a mountaintop, where you're closer to space and starlight has to pass through less air to reach you. But the first high-altitude observatory wasn't built until 1888. The money came from an eccentric American called James Lick, who made his fortune making and selling pianos in South America. He wanted a splendid memorial, such as a giant pyramid in the middle of San Francisco. Friends suggested that the world's biggest refracting telescope would be more highly appreciated, and that downtown San Francisco would not be the best location.

The Lick Observatory was instead built on top of Mount Hamilton, 1,283 m above the region now called Silicon Valley. Since then, every major observatory has been built high up. The current record-holder is the University of Tokyo's Atacama Observatory, in Chile, at an altitude of 5,640 m — two-thirds the height of Mount Everest.

Though it's been overtaken in altitude, Lick Observatory is unique in one respect. When you observe with the giant refractor, you are never alone. At the base of the massive column holding up the telescope you'll find a plaque. It reads: "Here lies the body of James Lick."

The observatory on La Palma in the Canary Islands is above 2,400 meters of churning atmosphere.

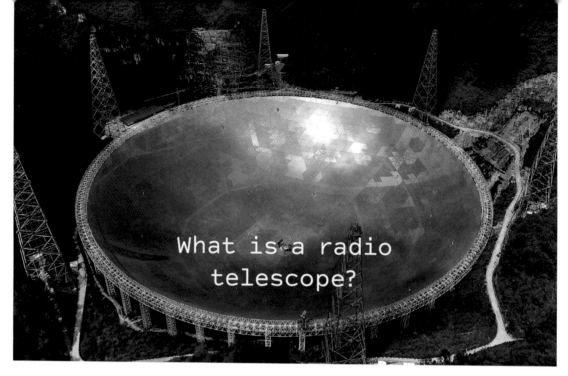

What is a radio telescope?

In 1931, engineer Karl Jansky investigated "static" that was interfering with the first trans-Atlantic telephone calls. He was surprised to discover these radio waves were coming from the Milky Way Galaxy.

At first, the world's astronomers paid no attention — they were preoccupied with the light from stars and galaxies. Only after World War II did researchers realize that cosmic radio waves came from places of extreme violence, such as exploding stars, and so were opening a whole new window to the cosmos. And they needed special telescopes to track down this radiation.

A radio telescope is simply a big aerial that picks up radio waves from space, exactly as a satellite dish receives radio signals carrying your favorite TV program from a satellite orbiting the Earth.

Unlike TV broadcasts, cosmic radio waves are very weak. So it helps to have the biggest telescope you can. Hence the giant dishes at Jodrell Bank in the UK, Arecibo in Puerto Rico and — opened in 2016 — the world's largest radio telescope, in China (see photo).

To find out what a cosmic radio source looks like, astronomers link up a whole array of radio telescopes and analyse the results by computer. Nigel was a radio astronomer at Cambridge and was lucky enough to use the Five Kilometer Telescope (eight dishes spread along the old Oxford to Cambridge railway line) to view the giant clouds of energy mushrooming out from quasars (see page 222).

Today, arrays are much bigger. The largest is the amazing Square Kilometre Array being built in Australia and southern Africa. The "square kilometer" refers to the area of all the individual telescopes if you add them together. However, the antennae cover huge regions of desert, and linking them electronically effectively makes a radio telescope that's thousands of kilometers across.

The Five-hundred-meter Aperture Spherical Telescope in southern China is the world's biggest radio dish.

How many different kinds of telescopes are there?

R adio waves and light are only two of the messengers bringing us information about the distant universe. If we only paid attention to them, it would be like listening to music with ears that can hear only the bass notes (radio waves) and the notes around middle C (visible light). To fully appreciate the symphony of the universe, we need telescopes that can tune in to other kinds of radiation.

An infrared telescope picks up wavelengths between radio waves and light, which are emitted by embryonic stars and galaxies. Earth's atmosphere tends to absorb this radiation, so you'll find infrared telescopes on high mountaintops, or better still, out in space — like the forthcoming James Webb Space Telescope.

Wavelengths shorter than light are absorbed by the atmosphere too, so astronomers must send ultraviolet telescopes into space to detect them. They observe hot stars and quasars.

X-rays have an even shorter wavelength, emitted by superhot gas at a temperature of millions of degrees. X-ray telescopes reflect radiation not with a curved dish but with the polished interior of a tapering cylinder. They reveal gas pools within galaxy clusters and the last shrieks of matter as it spirals into black holes.

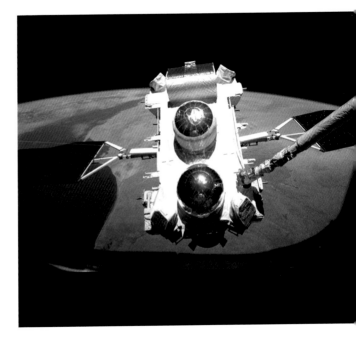

At the very top end of the cosmic keyboard are gamma rays. A gamma ray telescope, like the Compton Gamma Ray Observatory in the picture, doesn't have a mirror at all, but detects this high-energy radiation directly — like a Geiger counter. With these instruments, astronomers can examine the most extreme events in the universe, like the birth of a black hole.

The Compton Observatory is released into orbit from the space shuttle to observe high-energy gamma rays from space.

What's the world's biggest telescope?

There's an international race to build the biggest telescope the world has ever seen — watch the news over the next few years to see the winner emerge. Because a bigger mirror collects more light, the prize will be the first views of the faintest, and therefore the most distant, objects in the universe.

The one to beat is the current record-holder, the Gran Telescopio Canarias. Perched on the edge of a volcanic crater in La Palma, Canary Islands, it has a main mirror that's 10.4 m across. You can't make a single piece of glass that size and transport it to a remote mountain peak. Instead, it's a mosaic of 36 carefully shaped mirror panels fitting together snugly to make up a continuous curved reflecting surface.

But the Gran Telescopio Canarias is about to lose its crown. Next in line is the Giant Magellan Telescope, located in Chile. Its seven giant mirrors, nestled side by side in the same frame, will direct light to a single focus. Altogether, it will be as powerful as a single mirror 24.4 m across. Currently under construction, the Giant Magellan Telescope should open its eyes to the sky in 2021.

The Thirty Meter Telescope does what it says on the can: its 30-m-wide mirror is a big brother of the Gran Telescopio Canarias, made from an incredible 492 individual mirror panels. This long-awaited instrument has fallen behind schedule because of disputes as to where the beauty should live. First choice is the summit of Mauna Kea on the Big Island of Hawaii, the location of a clutch of other major telescopes. But — after the ground-blessing ceremony in 2014, construction was put on hold because Native Hawaiians claimed it would desecrate sacred land. Another option would be La Palma, next to the Gran Telescopio Canarias.

But the telescope that will knock out all of the opposition is the European Extremely Large Telescope, already under construction on a mountaintop in Chile and due to open its superlative eye in 2024. The main mirror, made of 798 panels, will be 39.3 m across, with the area of almost four tennis courts! The E-ELT will explore the faintest objects in the distant universe for the very first time; it will also examine planets of nearby stars for signs of life . . .

The forthcoming European Extremely Large Telescope (seen in this computer-generated model) will gather almost as much light as all existing telescopes put together.

How far can the most powerful telescope see?

From its perch in orbit around the Earth, the Hubble Space Telescope has a uniquely clear view of the distant cosmos. The longer the exposure time its camera is set to, the farther it can see.

To snap the incredible picture above, Hubble stared at the same small patch of sky in the constellation Fornax (the Furnace) for three weeks. Almost everything here is a galaxy way beyond our Milky Way. The faintest objects in this Hubble Ultra-Deep Field view are one ten-billionth the brightness that the human eye can perceive, and they lie 13.2 billion light years away.

Their light has taken so long to reach us that we're seeing these distant galaxies when they were very young, soon after the Big Bang that created the universe.

The new James Webb Space Telescope will peer even farther into the depths of the cosmos. But even its gaze will be limited to 13.8 billion light years. That far out in space, we are looking right back to the time of the Big Bang itself — and we can't ever see beyond that!

The ultimate telephoto shot — pull focus to 13 billion light years.

Why do astronomers fly telescopes in space?

In 1946 — just four years after the first rocket reached space and a decade before the first satellite — American astronomer Lyman Spitzer dreamed a dream. What if, he mused, we could put a telescope not just on a mountaintop, but totally above the Earth's veil of air. The light reaching such a "space telescope" would be unsullied by turbulence in the atmosphere; it would have a crystal-clear view of the cosmos.

The Hubble Space Telescope has measured the age of the universe.

Spitzer's vision eventually came to fruition 44 years later when space shuttle *Discovery* lofted the Hubble Space Telescope into orbit. Astronomers looked forward to highly detailed views of stars and galaxies, but all that the telescope sent back were blurred views. Oops, how embarrassing for NASA — the manufacturers had polished Hubble's mirror to the wrong shape . . . Fortunately, Hubble is near enough that astronauts could reach it and install corrective lenses.

For almost 30 years, the telescope has lived Spitzer's dream, sending us uniquely detailed images of planets, nebulae and galaxies near the edge of the observable universe. Hubble has revealed planets orbiting other stars and supermassive black holes lurking in galaxies. It has measured the age of the universe and discovered that the cosmos is accelerating — blowing up out of control. Hubble's view of the Pillars of Creation (see page 178) has become the iconic image of space for a whole generation.

And astronomers are expecting a new quantum leap with NASA's new eye in space, the James Webb Space Telescope, named after the administrator who steered NASA through the heady 1960s. Webb has a mirror that unfolds after launch, like petals opening up into a reflector flower with five times the area of Hubble's mirror.

Webb is the ultimate telescope. It will probe the first galaxies to light up the universe, probe into the dark nurseries where stars and planets are being born in our Galaxy, and investigate new worlds — in our Solar System and beyond — where alien life may be thriving.

James Webb Space Telescope: the future of astronomy

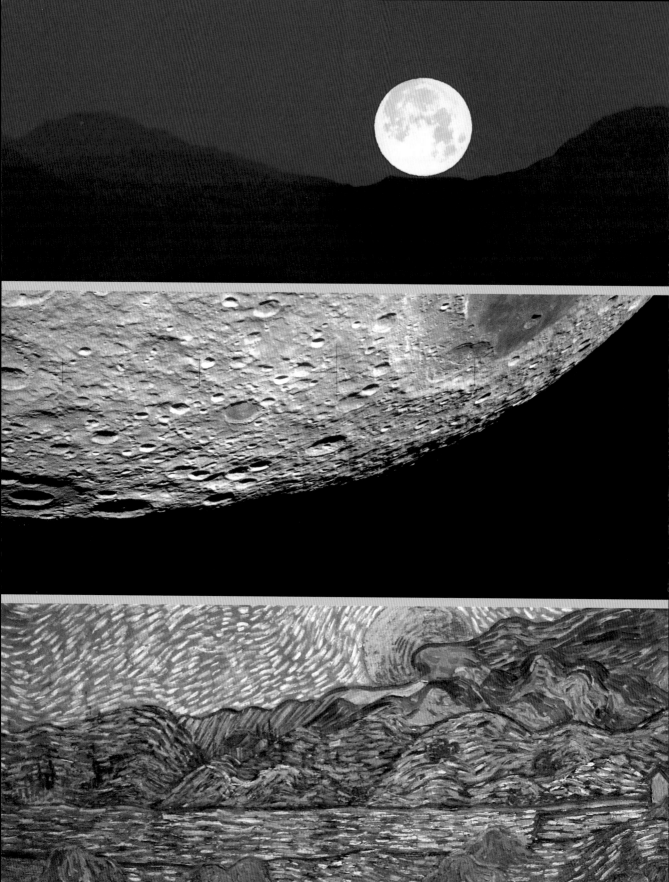

THE MOON 5

Did astronauts really land on the Moon?

Yes! There's no doubt that 12 astronauts on six missions did walk on the Moon between 1969 and 1972. The only reason we have included this question is that some people are still misled by the phony "evidence" touted by conspiracy theorists, who allege the missions were filmed in a movie studio on Earth.

None of their claims stands up to even the slightest scrutiny. For instance, why does the American flag wave if there's no air on the Moon? The flag initially wobbles because the flagpole is being moved.

Why aren't there any stars visible? The Moon's surface, and especially the astronauts' suits, are very bright, so the cameras were set to minimum sensitivity and the stars were too faint to register. The shadows were not completely dark because the lunar surface, lit brightly by the Sun, was reflecting light back into the shadows.

The technical arguments fare no better. The Van Allen belts posed little danger to the *Apollo* astronauts. Why did the video of the astronauts taking off not show a flame from the rocket engine? Because the fuels used — hydrazine and dinitrogen tetroxide — don't produce any visible flame when they burn.

When you start to look at the wider context, the conspiracy theory is even more laughable. Is it really possible that the U.S. government was able to silence every single NASA employee then, and for the following 40 years? Think of the current wave of leaks about even more politically sensitive issues.

The Russian intelligence service could certainly have found out if the Moon missions were a hoax, and they would have had the most to gain from exposing it publically. Clearly they knew the landings were genuine.

Also, the TV pictures from the first spacewalk were picked up by the Parkes radio telescope in Australia. When we visited there recently, the Australian astronomers (surely not in the pay of the CIA!) assured us that the telescope was pointing at the Moon, not at a studio in America.

We've been privileged to know many of the *Apollo* astronauts over the years. They were not liars but brave men who risked their lives to advance the frontiers of cosmic discovery.

Apollo 17 Commander Eugene A. Cernant taking a stroll on the Moon's surface

Where did the Moon come from?

The young Solar System was a place of incomparable violence (see Chapter 7). Worlds collided, asteroids scarred infant planets and others were utterly destroyed.

The Moon rocks collected by the *Apollo* astronauts have revealed the astonishing story of how this cosmic battleground brought our satellite into being. The composition of many of the lunar rocks is virtually identical to rocks on our planet.

This leads to one conclusion: that the Moon was blasted out of the Earth.

The "Big Splash" was a cataclysm that befell our world over four billion years ago. A wayward planet headed our way — hell-bent on destroying the young Earth. The Mars-sized body hit its target, almost destroying our fledgling world. But the larger Earth held its ground, pulling itself back together as a globe of molten lava.

But the impact left a permanent legacy. Matter from the impact splashed into space, forming a fiery, incandescent ring around our planet. The droplets then came together to create the Moon.

When you look at the tranquil face of the Moon today, it's hard to believe that it had such a traumatic birth — in fire.

The Big Splash: creation of the Moon

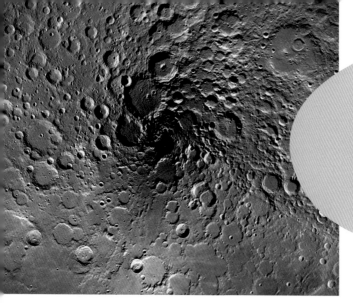

What made the Moon so cratered?

When Galileo pointed his telescope at the Moon, he was astonished to find that its surface was peppered with holes. He called them craters, after the Greek word for bowl.

What caused them? The answer isn't extinct volcanoes (which was one of the popular theories going around in the 1940s).

The Moon has thousands of craters. Most of them date from a period of bombardment billions of years ago.

The return of over 380 kg of Moon rocks collected by the *Apollo* astronauts revealed the truth: the Moon's craters are testimony to its violent history. It was mercilessly struck by asteroids and meteorites in its youth, when our Solar System was forming. The Earth, too, suffered; but its dense atmosphere and active geology has largely erased the scars.

Not so on the inactive, airless Moon. It has thousands of craters, all perfectly preserved. Most of them date from a period of bombardment billions of years ago.

Once craters were discovered, astronomers embarked on what they do best: giving them names. In 1651, Giovanni Battista Riccioli named 247 craters, immortalizing eminent scientists and philosophers — including himself!

Craters are complex. The impact first makes them cave in, and then rebound, creating a central mountain peak. The walls of the crater then collapse to form a series of terraces.

Not all craters are ancient. Magnificent Copernicus, 93 km across, was blasted out 800 million years ago. And stunning Tycho (102 km across) is a mere 108 million years old. Both these craters are surrounded by colossal ray systems of ejected rock.

Best time to see craters? When the Moon is half lit and the sunlight comes side-on. They're a sensational sight, even through the smallest of telescopes.

The pockmarked north pole is typical of the Moon's highly cratered surface.

What caused the face of the Man in the Moon?

This magnificent image is a frame from a futuristic French film from 1902 called *Le Voyage dans la Lune*. We're not suggesting that a mighty spacecraft created the features of the Man in the Moon (see Chapter 2), but the filmmakers were on the right track.

As we've seen in the last question, the Moon's craters were formed by intense bombardment by asteroids and meteoroids during the early years of the Solar System. It culminated 3.8 billion years ago in the mother of all batterings: the Late Heavy Bombardment.

Giant asteroids struck the Moon, gouging out enormous basins; they're far too large to be called craters. Most are bigger than 700 km across; the largest — Oceanus Procellarum — is 2,568 km in diameter.

The impacts penetrated deep inside the Moon itself. Magma welled up into the basins, creating smooth, dark surfaces — the familiar face of the Man in the Moon (or Hare, if you prefer).

The early telescopic astronomers — no doubt inspired by the sea voyages at the time — believed that these basins were seas, or *maria*. The names have stuck: Mare Imbrium (Sea of Showers); Mare Tranquillitatis (Sea of Tranquility); Mare Frigoris (Sea of Cold).

When NASA was planning its *Apollo* landings, it recognized the maria for what they were: potentially smooth landing sites. *Apollo 11* landed in Mare Tranquillitatis, and *Apollo 12* in Oceanus Procellarum, the biggest of the seas. Later missions ventured farther afield.

And it's still in the cards that we may ourselves roam these lunar seas again — this time in search of a permanent habitation.

In the pioneering color science fiction film *Le Voyage dans la Lune (A Trip to the Moon)*, the Man in the Moon suffers a distinctly man-made impact.

The Moon is a lazy man called Ngalindi. We see him big and round and fat in the sky at the time of Full Moon. But his wives are resentful, and gradually hack bits off Ngalindi, so that he shrinks in size. Eventually, he dies and disappears at New Moon. After three days, Ngalindi rises again and grows fatter and lazier — until at Full Moon his wives attack him again.

That's how the Yolngu people of northern Australia explain the changing shape of the Moon night by night during each month (an English word, by the way, that was originally "moonth").

> When we look at the Moon in the sky, we can see only the sunlit part of the lunar globe — and the amount changes as the Moon moves around its orbit.

It's an engrossing story, but not how astronomers explain the changing phases of the Moon. Scientifically, it's more prosaic — it happens because the Moon has no light of its own.

The Sun lights it up, illuminating just the half that's facing toward the Sun. When we look at the Moon in the sky, we can see only the sunlit part of the lunar globe — and the amount of lit surface that we view changes as the Moon moves around its orbit.

When the Moon lies in the direction of the Sun, we are looking at the unlit portion, and we can't see the Moon at all — this is the New Moon. A couple of days later, as the Moon moves along its orbit, we can see a narrow region of the lit-up hemisphere, appearing as a beautiful thin crescent in the evening sky.

A week after New Moon, our celestial companion has gone a quarter of the way around its orbit — this is the time of the First Quarter, when we see half the Moon lit up. The Moon then appears to increase in size — in traditional terms, it "waxes" — through the bulging phase called "gibbous" until the Moon is opposite the Sun. All of its illuminated face is turned toward us, and the Full Moon appears as a complete glowing disc.

After Full Moon, we see less and less of the illuminated side, and the Moon appears to shrink (or "wane") in the morning sky, past the Last Quarter and crescent phases to the next New Moon.

The phases of the Moon result from the changing illumination of sunlight (coming from the right, here) as the Moon orbits the Earth.

First Quarter

Waxing Gibbous

Waxing Crescent

Full

New

Waning Gibbous

Waning Crescent

Third Quarter

What is the Harvest Moon?

The helpful Harvest Moon is the Full Moon closest to the autumn equinox — in September in the Northern Hemisphere. Any Full Moon lights up the countryside, but there's something special about the Harvest Moon.

Throughout history, farmers have fought against the shortening daylight as they reap the golden reward of their summer toil. And, before the days of electric lights, they had a brilliant ally in the sky. In his legendary poem "The Lady of Shalott," Alfred, Lord Tennyson, describes how "by the Moon the reaper weary" piles his sheaves of corn "in uplands airy."

Around the time of the autumn equinox, the Moon's monthly path through the sky lies at a shallow angle to the horizon. As a result, it rises only a little later on each successive night, and for several evenings farmers are blessed with a bright Moon rising soon after the Sun has set.

Painting by Van Gogh, inspired by the Harvest Moon rising over wheat sheaves

Why does the Moon look so big in June?

The Full Moon in June is guaranteed to stir your heart — that great silvery-gold orb hanging low on the horizon as you promenade through the warm summer evenings with your love on your arm. (Except in the Southern Hemisphere, where the Moon in December has the same effect but not the same poetry.)

But the scientific reason for the giant summer Moon is rather more mundane. First, around midsummer you'll always see the lowest Full Moon of the year, because a Full Moon lies opposite the Sun in the sky (see page 96), and the Sun is at its highest in June.

And when the Moon lies near the horizon, it invariably looks bigger. But this is pure illusion. The Moon is not any closer to us. And it's nothing to do with the way moonlight passes through Earth's atmosphere. Although the atmosphere *does* change the Moon's color to a pale gold (just as it reddens the setting Sun — see page 13), it has no effect on the Moon's apparent size.

The effect has been known for millennia. Greek astronomers first noted it, and illustrious astronomers through the ages have written about the phenomenon. But it's not a physical effect; it's a trick played by the human eye. When you see the Moon low against a horizon of trees and houses, your mind compares it to these nearby objects — and so it appears larger. You can prove this by photographing the Moon at hourly intervals from rising to setting, then measuring the images. We guarantee that there'll be no change in size.

How to make the illusion go away? It's rumored that standing with your back to the Moon, bending down, and looking at it between your legs will make it appear normal again. Just don't let your friends catch you in the act . . .

Hovering on the skyline of the Mojave Desert, the Full Moon looks deceptively large.

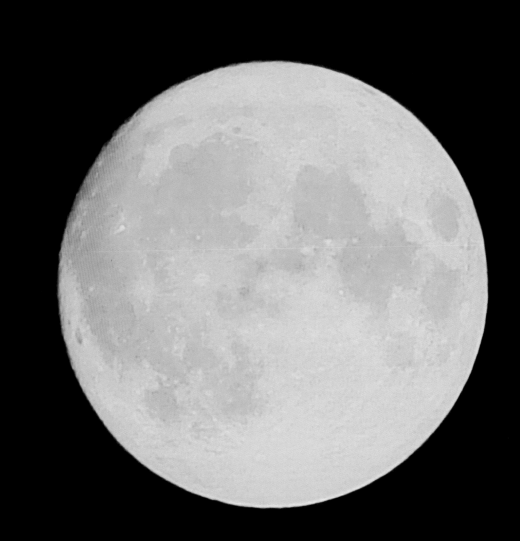

What is a Blue Moon?

"I only see you once in a Blue Moon," you may complain to a friend. The concept dates way back to medieval times, when the phrase "the Moon is blue" meant something impossible — though it mellowed over time to refer to events that are uncommon. But how rare is a Blue Moon?

Well, it all depends on what you mean. The Moon is silvery white when it's high in the sky, and often pink or red near the horizon (as we mentioned in the previous question). But in September 1950, people around the Northern Hemisphere were amazed that the Moon appeared a brilliant blue. The culprit was a vast forest fire in Canada. It wafted specks of ash high into the air that were just the right size to obstruct the long wavelengths of red and orange light. As a result, the moonlight filtering down to the world below was predominantly blue.

The eruption of Krakatoa in 1883 caused Blue Moons, and also — to a lesser extent — the ash from Mount St. Helens (1980) and Pinatubo (1991). But they are very rare. According to this definition, you'd be lucky to see your Blue Moon friend two or three times in a lifetime.

But surely you've heard the phrase mentioned more often than that? Yes indeed. There's another definition that stems from old farmer's almanacs. In North America, the 12 Full Moons in a year have special names, such as the Harvest Moon in September and the Wolf Moon in January. Occasionally, though, there are 13 Full Moons in a year. Because the extra Full Moon was uncommon, the almanac writers called it the Blue Moon.

In 1946, a writer for *Sky & Telescope* magazine misinterpreted what was going on. When two Full Moons fall in the same month, he declared, the second one is called a Blue Moon. Forty years later, the Trivial Pursuit game included this definition as the answer to their question "What is a Blue Moon?" And so the mistaken definition became the norm that almost everyone uses today.

If your definition is the second Full Moon in a month, then you can get more social — look forward to meeting up with your Blue Moon buddy once every two-and-a-half years.

Don't expect a "Blue Moon" to resemble this touched up image. In reality, it looks just like any other Full Moon.

The short answer is, emphatically, NO. The Moon — in all its phases — does not influence human behavior.

But hang on, what about the synchronicity between the length of the lunar phase cycle and the regularity of women's periods — both roughly 28 days? A complete coincidence, say biologists. Other mammals' fertility cycles range between 4 days (rats) and 60 days (dogs); humans happen to come out somewhere in the middle.

Coming to the Full Moon: there are apocryphal stories from police officers that more crimes take place when the Moon is full, and nurses report more admissions into the emergency room. Neither claim has stood up to scrutiny.

There have been hundreds of clinical investigations into

Werewolf having a bad night

the correlation between erratic human behavior and the Full Moon. After all, the word "lunatic" comes from Luna, the Roman goddess of the Moon. Both Greek and Roman philosophers believed that the Moon could disturb the waters in the brain when it was Full; but tides only ebb and flow in open bodies of water (see page 106).

> The word "lunatic" comes from Luna, the Roman goddess of the Moon.

Scientists have also looked into ways that the Full Moon affects birth rates and epilepsy. Again — no link. And an investigation into 5,812 children of all nationalities studied sleep deprivation at Full Moon. The average was five minutes less slumber — hardly life-threatening!

Now for the fun part. What about were-wolves appearing at Full Moon? After all, there have been some spine-chilling werewolf movies recently. And in medieval Europe, the cult of the werewolf (along with witches) was rampant. This ingrained belief holds quite strongly even today. In central Europe, wolves were the most feared predator. And it was even more terrifying to contemplate an ordinary human being shapeshifting into a wolf, with all its gory implications.

But how many werewolves have you seen? Exactly. Yet, they may have an unexpected cousin: wading birds. Their hunting cycles peak at the Full Moon — not because of the extra light, but as a result of the tides. At Full Moon, the Sun, Earth and Moon are in line, and the extra pull of the Sun makes the tides slightly higher.

Experts are now studying why and how this hunting behavior takes place. In the meantime — avoid the werewolf wading birds!

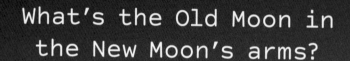

What's the Old Moon in the New Moon's arms?

You're sure to have seen this beautiful sight: a brilliant young crescent Moon wrapped around the ghostly globe of the Old Moon. You can still see the Old Moon's details; but they're faint. It's a delightful old description of a lovely phenomenon, but what causes it?

The answer is Earthshine — reflected sunlight from the Earth lighting up the nighttime Moon. You don't get to see it every time the Moon is in the sky, because a lot depends on the Earth's weather. Oceans rule our planet's brightness when the weather's clear, but water does not reflect sunlight well. Clouds, on the other hand, can bounce back up to 50 percent of the Sun's rays.

So brighter Earthshine means a cloudy planet. In fact, some scientists monitor Earth's cloudiness by observing Earthshine on the Moon, which in turn provides a guide to the development of the greenhouse effect.

A slender crescent Moon embraces the fading Old Moon.

What is an eclipse of the Moon?

Ever seen the Moon looking a bit unwell? Dim and copper-colored? Then you're probably witnessing a lunar eclipse.

They take place at Full Moon, when the Sun, Earth and Moon are in a straight line. But they don't take place every Full Moon. Our natural satellite orbits the Earth at an angle, allowing the Sun's rays to pass over or under the Earth, bathing the Moon in its light. But sometimes the alignment is exact. The Moon passes into Earth's shadow, sunlight is cut off, and the result is a lunar eclipse.

The ancient Greeks knew the cause of lunar eclipses. They also observed that when Earth's shadow started to engulf the Moon, it was curved — proving that our planet is round. From the size of the shadow, they deduced that the Earth was four times bigger than the Moon.

Eclipses of the Moon aren't rare: there are usually at least a couple per year. And they can be seen by half the world at once — unlike solar eclipses (see Chapter 6), where the alignment between Sun, Moon and Earth is critical.

What to expect? First, there is *no* danger in watching a lunar eclipse; the Moon simply drifts into the Earth's shadow and out again. This can take up to four hours. Totality is reached when the Moon enters the deepest part of our planet's shadow. This can last for over an hour.

The eclipsed Moon seldom disappears. Sunlight gets refracted onto the Moon by the Earth's atmosphere, causing the Moon to shine with a dull reddish glow. The more dust there is in the atmosphere (for example, from volcanic eruptions), the darker the eclipse will be.

Do animals react? In our experience, yes. We were in Cartagena, Colombia, on a lecture tour. There was a lunar eclipse, which we watched from a balcony overlooking a pool. Below, a bullfrog was merrily chirping.

Come totality, he fell completely silent. An hour later — when our friend saw the first glimmer of the uneclipsed Moon emerge — he recommended his performance. A lovely demonstration that animals, too, are aware of the night sky.

The eclipsed Moon appears coppery, lit by sunlight reddened and refracted by Earth's atmosphere.

How does the Moon cause the tides?

Around AD 1030, the mighty Cnut — King of England, Denmark and Norway — set his throne on a beach and commanded the advancing tide "not to rise on to my land, nor to presume to wet the clothing or limbs of your master." But the sea paid no heed. Jumping back with his legs soaked, Cnut cried out, "The power of kings is empty and worthless, and there is no king worthy of the name save Him by whose will Heaven, Earth and sea obey eternal laws."

Though he didn't know it at the time, the "eternal law" in question is gravity. And, when it comes to the tides, we're mainly talking about the Moon's gravity.

On the side of the Earth that's facing the Moon, our companion's gravity pulls the oceans upward, producing a high tide. There's a simultaneous high tide on the opposite side of the globe, because the Earth is swinging around the balance point of the Earth-Moon system and the water is flung outward.

As the Earth spins around on its axis, every seaport and beach on the planet is carried through these "tidal bulges" twice a day, with low tides in between. If you live by the sea, you'll know that the tides don't repeat every 12 hours, but in a period of 12 hours 25 minutes. That's because the Moon has moved part of the way around its orbit by the time we come to experience the next tide.

The Sun's gravity also pulls on the oceans. When the Sun and Moon are in line with the Earth — at New Moon or Full Moon — it reinforces the Moon's influence, and we experience "spring tides," which are more extreme than usual.

Some people seem convinced that the Moon's tidal influence must affect us because our bodies are largely made of water. But remember that the tides in the ocean are tiny compared to the size of the Earth, and the Moon raises tides in our bodies that are only one-hundredth the size of a body cell.

Portishead, on England's Severn Estuary, has one of the world's greatest tidal ranges.

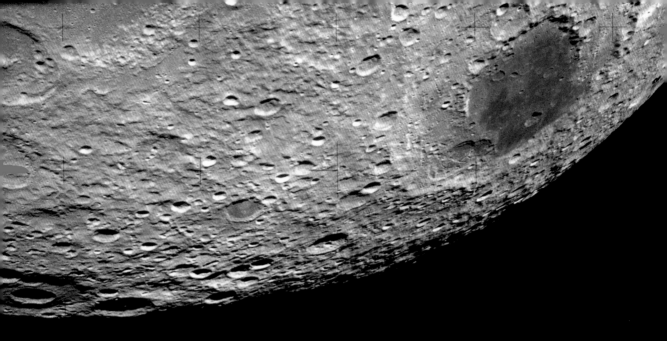

What is the "dark side" of the Moon?

There *isn't* a "dark side" of the Moon! It's the most common misconception in astronomy that because the Moon always has one hemisphere turned toward the Earth, its back side must be shrouded in darkness.

Not so. As the Moon orbits the Earth, *all* of its surface catches the Sun. The confusion stems from the fact that gravity has "braked" the Moon's spin. But because of the Moon's slightly oval orbit, we can get tantalizing glimpses of the Moon's far side around the edge of its familiar face.

Our first view of the 41 percent of the Moon we can't see from Earth came in 1959, when Russia's *Luna 3* flew past the far side and returned photographs. Unlike the hemisphere that faces us, the lunar far side lacks the huge impact basins we're familiar with. But it's covered in craters, packed together side by side.

In 1968, the crew of *Apollo 8* were the first humans to witness the far side. Lunar Module pilot Bill Anders came up with a great description: "The back side looks like a sand pile my kids have played in . . . It's all beat up, no definition, just a lot of bumps and holes."

The far side of the Moon, lit by the Sun — often mistakenly referred to as the "dark side."

Can I buy a Moon rock?

Pete Conrad, third man on the Moon, looks as if he's about to walk off with a lovely pair of dark stones that he collected from the lunar surface. In fact, the astronauts have no rights over their spoils. NASA jealously guards the one-third of a ton of rocks that the *Apollo* astronauts brought home, preserving these precious stones for scientific research.

Apart, that is, from the first small bag of rocks and soil that Neil Armstrong collected on the *Apollo 11* lunar mission. Somehow, this ended up in the possession of the manager of a museum in Kansas. When he was convicted of theft, U.S. Marshals seized the bag. Later they put it up for auction, and a Chicago-area attorney, Nancy Carlson, bought it for $995.

Carlson sent the bag to NASA to check its authenticity, but the space agency hung onto it, claiming the lunar samples as its own property. Eventually, a U.S. district judge ruled in Carlson's favor, and NASA had to return it to her. In July 2017, Sotheby's auctioned the bag on Carlson's behalf — for $1.8 million.

Don't worry, there are ways to get your hands on a Moon rock more cheaply (and we don't mean flying to the Moon yourself!). Small asteroids are always blasting rocks out of the Moon, and some of these fall to Earth as meteorites — over 300 have been collected so far, the biggest weighing a whopping 13 kg.

Meteorites are most easily seen in desert landscapes, and the Sahara is an ideal hunting ground. Local people in northern Africa sell the lunar rocks they find to meteorite dealers, who cut up the rocks and sell fragments on the open market. Search online for "lunar meteorites for sale," and you can possess your own tiny piece of the Moon for less than $50!

Apollo astronaut Pete Conrad handles his Moon rocks.

Can people live permanently on the Moon?

Yes, and we know where: on one of the Moon's romantically named Peaks of Eternal Light.

Near both the Moon's poles, there are a few mountains so high that they are bathed in sunlight almost continuously as the Moon rotates. That's ideal for a lunar base, as the colonists would enjoy a nearly uninterrupted supply of solar power. Another advantage is that the temperature at these peaks stays nearly constant at −50°C (−60°F), similar to an arctic winter — unlike the Moon's equator, which swings from a scorching 130°C (260°F) during the day to a freezing −170°C (−280°F) at night.

The Moon's north pole lies on the edge of a crater named for American explorer Robert Peary, who claimed to be the first to reach the Earth's North Pole. The massive peaks on the rim of Peary crater are bathed in sunshine over 90 percent of the time.

Or you could head south. The Moon's other pole lies within a crater that bears the name of Ernest Shackleton, the British Antarctic explorer who paved the way for Raoul Amundsen and Robert Falcon Scott to reach the Earth's South Pole. Peaks near Shackleton crater are almost as sun-drenched as the rim of Peary crater.

There's a bonanza hidden deep in Shackleton crater, forever shaded from sunlight — the temperature never rises above −180°C (−290°F). Over billions of years, water vapor drifted over

Proposed European base on a Peak of Eternal Light

the Moon each time an icy comet impacted, and some of the water has condensed and frozen at the bottom of Shackleton. According to measurements from NASA's *Lunar Prospector* spacecraft, ice may make up one-fifth of the soil down there. Future colonists perched on the crater's bright rim will be able to mine this ice to create hydrogen and oxygen, for fuel and for breathing.

Will the Moon collide with the Earth?

The sky is falling, the sky is falling!" So Chicken Little exclaimed to the world after an acorn fell on her head. As astronomers, we're pretty sure the sky isn't really going to descend — but, one day, the Moon *will* fall to Earth. We're glad to say it's a pretty long time away . . .

At the moment, the Moon is actually moving away from the Earth. The *Apollo* astronauts left retro-reflectors on the Moon (like the reflectors on the back of your car), and the Russians mounted similar reflectors on the *Lunokhod* lunar rovers. By bouncing laser beams off them, scientists can measure the Moon's distance with phenomenal accuracy. They've found the Moon is retreating from us at 38 mm per year. Since Neil Armstrong made his "one small step," the Moon has moved away by about his height.

The Moon's retreat is caused by those same tides that the Moon generates on Earth. These bulges of water are pulling ever-so-slightly on the Moon, making it move gradually forward in its orbit, and so propelling it outward.

Some five billion years from now, the Sun will expand into a red giant star. It may grow so big that its outer gases lick the Earth and Moon, slowing down the Moon's orbit and possibly putting it in a death-spiral toward the Earth.

Even if the Sun spares us, the Moon won't travel away forever. When its orbit stretches to 47 days (rather than today's 27 days), tides will have braked the Earth's rotation so that our "day" is also 47 of our current days. This means that one side of the Earth always faces the Moon, just as the Moon always has the same hemisphere turned toward our planet. Now the Moon begins to move back toward the Earth. And 65 billion years from now, it's poised to strike.

Whether it happens sooner or later, the outcome will be the same. As the Moon comes into low orbit around the Earth, our planet's gravity tears it apart. The Moon disintegrates above the planet to form a set of rings — like Saturn's — around the Earth. And then the final act. These giant rocks that once formed the Moon crash down to Earth as giant meteorites. They blaze through the atmosphere and smash into the ground, melting it into a giant ocean of incandescence.

The Earth and Moon as seen from NASA's Galileo spacecraft

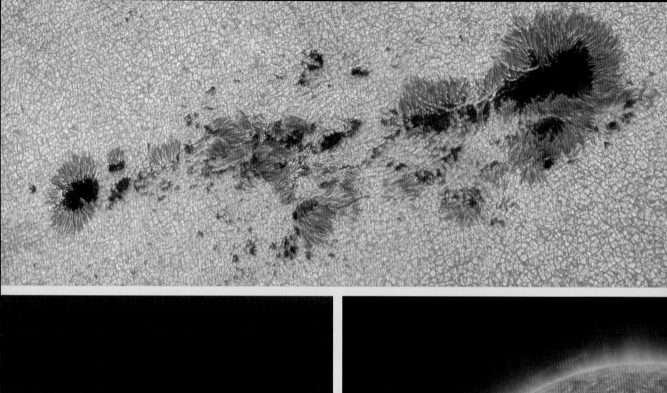

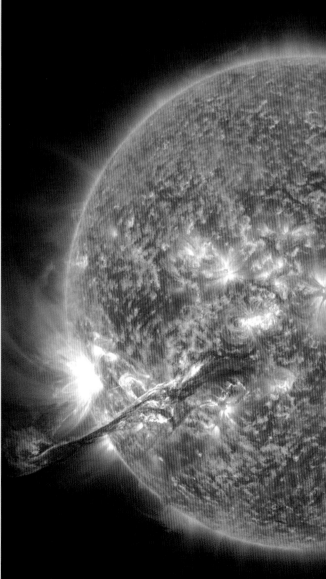

THE SUN 6

Is the Sun a star?

Heather was pitching for a TV series — *The Stars* — to a well-educated commissioning editor at UK's Channel 4.

It was an uphill struggle, but she eventually relented. "Heather," she observed, "you said something very important today. You told me that the Sun is a star. Tell me . . . is the Moon a star?"

Gasp! But she got the series.

There is a world of difference between the Sun and the Moon. Although they both light our skies, the Moon shines by reflecting sunlight — just like all the moons and planets in our Solar System. But the Sun is a different kind of beast. It's self-luminous (see page 124) — a result of its

The Sun seen from Pluto — looking like nothing more than a star

mighty gravity and vast size. You could fit one *million* planets like the Earth inside the Sun.

The reason for the confusion? The Sun is on our doorstep, and it controls our Solar System. The other stars in the sky look like pinpricks in the firmament — but that's only because they're so far away. Many of them are bigger, hotter and more violent than the Sun (see Chapter 9).

Yes, the Sun *is* a star — our very own local star. And without its prolific energy, we wouldn't be here to write about it.

How long would it take to fly to the Sun by plane?

Like it or not, long-haul flights are a way of life these days. We think nothing of the ten-hour journey to London from Los Angeles. But let's put things in an astronomical context, and envision a trip to the Sun — by plane!

The average long-haul jet cruises at a speed of 980 km/h. We're going to fling in a couple of planets en route — the innermost worlds, Venus and Mercury. So, seat belts fastened, and off we go.

Venus is the nearest planet to Earth. At its closest, this cloud-covered enigma of a world can approach our planet to a distance of 38 million km. Travel time? 40,000 hours — that's four and a half years.

Next, it's pockmarked little Mercury, the closest planet to the Sun. It's 77 million km away from the Earth at its closest, and our jet would get there in just over nine years.

Now we're on our final lap — to the Sun itself. 150 million km distant, it lies at the center of its immense family of planets, comets, moons and asteroids. The ultimate journey would take a staggering 155,000 hours — almost 18 years. Now you just have to go through customs and immigration . . .

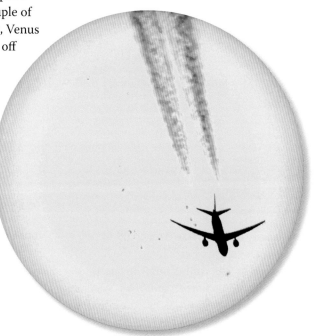

Is it safe to look directly at the Sun?

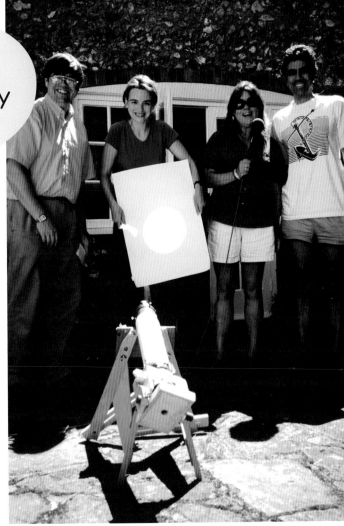

NO! You could damage your eyesight, and in the worst cases, risk blindness (see the next question).

We all know from sunbathing warnings that the Sun zaps us with ultraviolet rays. This searing, short-wavelength radiation isn't good news for human beings. Especially when you eyeball the Sun too much. It causes scarring in the retina, which can lead to blindness.

Even more dangerous is looking at the Sun through binoculars or a telescope. Here you concentrate the sunlight on the retina, with disastrous effects. Even looking at the reddened Sun when it's setting low on the horizon isn't safe: long-wavelength infrared rays can still get through — not good news for your eyes.

Rumour has it that Galileo became blind as a result of observing the Sun through his telescope. This isn't true; he developed glaucoma late in life, which led to his blindness.

Galileo observed the Sun in the safest possible way: by using his telescope "in reverse" and projecting the image onto a piece of card.

These days, there are methods of observing the Sun safely with optical aid. If you're a Sun fan, buy yourself some solar binoculars with filters built into the lenses. Take our word for it — the views are fantastic!

Heather with a radio production team, showing the safe way to observe the Sun: solar projection.

What is a total eclipse of the Sun?

Nothing prepares you for a total solar eclipse. It's one of the most awesome sights you will ever get to see. We've cached up seven — from locations as diverse as Indonesia, Tahiti, Egypt and the U.S. Even as experienced astronomers, we're struck by a sense of primal terror when the eclipse takes place.

Unlike lunar eclipses, which take place when our satellite drifts into the Earth's shadow, an eclipse of the Sun is an altogether different — and mindblowing — affair.

It all happens because of a remarkable coincidence. By a complete chance, the Moon and the Sun appear the same size in the sky. The Moon is 400 times smaller than the Sun — but our local star is 400 times farther away. So the Moon can exactly cover the Sun during a total solar eclipse.

Eclipses don't take place on every orbit of the Moon, because of its tilted orbit. And you have to be in exactly the right place on Earth to see the crucial overlap. As a result, total solar eclipses are much rarer than eclipses of the Moon — roughly once every 400 years in any given place.

So here's what to prepare yourself for. Check out when and where your next total eclipse is due by going to Wikipedia for their list of solar eclipses, which stretches from 2000 BC to AD 3000! Have your eclipse goggles ready for the next viewing party.

What to expect? A growing chunk is carved out of the Sun as the Moon advances across its disc. Then, as totality approaches, you'll feel a drop in temperature, dampness and perhaps a light wind. All around you, birds and animals are becoming quiet — preparing to sleep. The light takes on an eerie, flat quality, like a film set.

Then, totality! Only during a total eclipse can the goggles come off. It's safe to look at the apparition that has replaced our comforting,

Total eclipse of the Sun: a once-in-a-lifetime experience

SAFETY WARNING: Always use eclipse goggles when you're observing an eclipse. DO NOT USE FOGGED FILM, WELDER'S GLASSES, OR ANY OPTICAL AID. Only if you're an experienced solar observer can you use specially designed solar binoculars. NEVER look at a partial eclipse without proper eye protection.

dependable Sun. Now there's a black disc in the sky surrounded by the devastatingly beautiful tendrils of the solar corona — the Sun's outer atmosphere — stretching into space. Stars and planets appear in the darkened sky.

And, almost as quickly as it began, the eclipse is over. A beam of light appears from a gap in the Moon's mountains, forming a glorious "diamond ring." GOGGLES ON!

Eclipses are very brief, and they seem even briefer when you're watching. The best you can expect is for totality to last about seven minutes. The much-publicized eclipse of August 21, 2017, cut a swathe across the middle of the United States — but at best, it was just 2 minutes and 40 seconds long.

If you were lucky enough to catch that eclipse, you'll heed our last word of advice. Seeing one eclipse isn't enough. Solar eclipses are addictive, so start a travel budget now for the next one!

Is it hottest in summer because the Earth is closest to the Sun?

It might be tempting to reach for "yes" as the obvious answer. In summer, the Sun seems so hot and close, while in winter it's wan and appears remote from us. But, hang on, you don't need to be a world traveler to know that when it's summer in the Northern Hemisphere, countries south of the equator are experiencing winter.

The Earth is tipped up as it orbits the Sun — remember how school globes are tilted at an angle of 23½ degrees. In June, the Northern Hemisphere is angled toward the Sun. People there see the Sun higher in the sky, and its heat falls more directly on the ground, giving them the warmth of summer. At the same time, the Southern Hemisphere is tilted away, and countries there experience the low Sun and cold temperatures of winter.

Six months later, the opposite is true. The Northern Hemisphere is cold, while people in the southern regions of Earth enjoy a blazing hot Sun high above them.

Though the Earth's tilt is responsible for the changing seasons, a couple of other factors affect how hot the summer can be. The first is that our planet is closest to the Sun around January 3 — only 147 million km away, as compared to 152 million km in early July — so southern summers should be slightly warmer than summers in the Northern Hemisphere.

The second aspect affecting the summer heat is the fact that most of the continents lie in the Northern Hemisphere, while huge oceans dominate our planet south of the equator. Landmasses warm up more quickly than ocean water, so summer temperatures rise faster in the Northern Hemisphere. As a result, the average temperature of the Earth is warmest during the northern summer, when our planet is farthest from the Sun.

The ultimate heat wave: in California's Death Valley

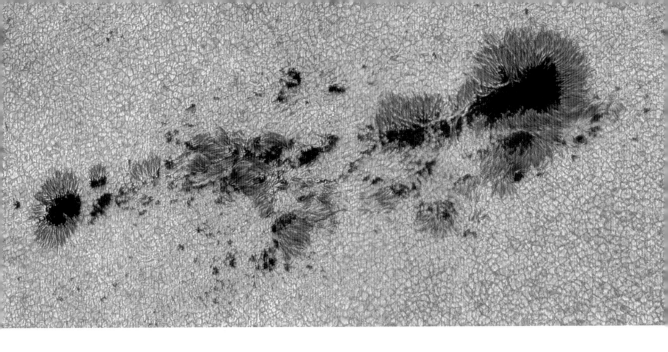

What is a sunspot?

A sunspot is a symptom of a stellar rash on our local star. The Chinese knew about them over 2,000 years ago; Galileo charted them too.

But what causes them? What follows is the best guess. Our Sun is made entirely of gas, with a surface temperature of 5,500°C (10,000°F). It spins faster at the equator than at the poles. And roughly every 11 years, our local star twists its magnetic field into the equivalent of a tight elastic band.

At sunspot maximum the Sun's magnetic field is squeezed into loops that restrict the motion of its turbulent gases. Where the magnetic loop breaks through the glowing surface, darker and cooler patches — sunspots — appear.

"Cooler" is relative. Although sunspots are darker than the Sun's surface, they are still at a scorching 4,000°C (7,200°F). If you could see a sunspot on its own, it would shine brighter than the Full Moon.

If you want to view sunspots yourself, check out the question "Is it safe to look directly at the Sun? " (page 119) If in any doubt, use solar projection onto a piece of white card.

Don't be disappointed if you can't see many spots at the moment. The Sun is near the minimum of its magnetic cycle, and weeks may go by without a spot appearing on its face. Hang on until the early 2020s, when the spots of a new cycle will be erupting on the Sun. Happy viewing!

A sunspot group eleven times wider than the Earth!

What makes the Sun shine?

"Nuclear Power? No Thanks!" This logo — invented by a pair of Danish antinuclear activists — has been around since the 1970s. It sports a cheerful, smiling Sun, giving the message that solar energy is natural and good, while nuclear energy is bad.

Well, it turns out that the Sun is actually a colossal nuclear reactor sitting right on our doorstep.

In the nineteenth century, astronomers had discovered how hot the Sun was and the rate at which it was releasing energy. Biologist Charles Darwin — in his controversial book *On the Origin of Species* — worked out that, to propel evolution, the Sun must be at least three hundred million years old. It was a staggering number in those days (although today we know the Sun has been around for 4,567 million years).

So what powers it? The American astronomer and aviation pioneer Samuel Pierpont Langley wondered if the Sun might be a giant lump of burning coal. But his calculations fell woefully short. Even if the whole Sun was made of coal, it would burn out in a few thousand years.

Over to distinguished physicist Lord William Kelvin in Glasgow. Inspired by the spectacular sight of incandescent shooting stars in our skies, he proposed that the Sun could be powered by an infall of meteors. That didn't stand up either. It would need a mass of meteors as great as the Earth impacting the Sun every 47 years. And that would have seriously upset the orbits of the planets.

It fell to Sir Arthur Eddington in the 1920s to come up with the goods. Just a few years before, a brilliant young British astronomer — Cecilia Payne-Gaposchkin — had discovered that stars are made almost entirely of hydrogen gas (see Chapter 9). Based on her discovery and Einstein's predictions of relativity (see Chapter 13), Eddington hypothesized that the high pressure and temperature at the center of the Sun could weld hydrogen atoms into the next-lightest element — helium — with the release of energy.

He was correct. In the Sun's core, nuclear fusion reactions convert four million tons of matter into energy *every second*. This is the energy that gives us light, that powers our lives, that gives us heat and that has provided the Earth with an abundance of life.

Think about it: without our local, slow-burning hydrogen bomb, you wouldn't be here!

Nuclear fusion reactions power the Sun and, on a much more modest scale, hydrogen bombs like Castle Romeo, tested here in 1954.

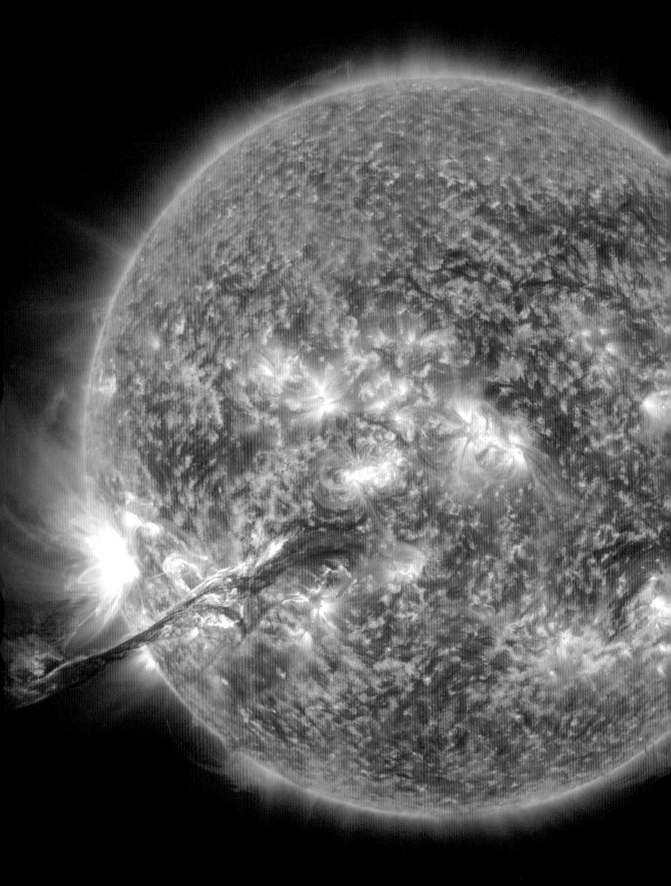

Could a solar storm wipe out life on Earth?

If there's a big, active sunspot group on the Sun's disc, look out for warnings of solar storms!

At sunspot maximum, the Sun is in a frenzy of magnetic activity. When loops of its magnetism touch above a spot group, they short-circuit and explode in a brilliant solar flare.

Solar storms disrupt satellite navigation systems and mobile phone networks, and some experts warn that a major shutdown of our electricity grids would lead to international mayhem.

Higher up in the Sun's atmosphere — in the pearly corona that you see during eclipses — there are even more dangerous eruptions called coronal mass ejections (CMEs). Both flares and CMEs eject streams of dangerous particles and radiation that wreak havoc in our Solar System.

In 1989, a solar storm wiped out the "fault-tolerant" computers at Toronto's stock exchange for three hours. It also hit Canada's national grid, cutting off power to six million people in Quebec.

Solar storms are inconvenient, but not fatal, to life on Earth. Okay, they disrupt satellite navigation systems and mobile phone networks, and some experts warn that a major shutdown of our electricity grids would lead to international mayhem. But our planet's magnetic cocoon is normally pretty good at protecting us by channeling the harmful electric particles from reaching the ground.

It's different in space. Away from Earth's magnetic shield, astronauts of the future — perhaps heading to Mars — will bear the full brunt of our Sun's unpredictable whims, a problem that space technologists are taking very seriously.

A solar storm: not good news for our planet

Will the Sun explode?

At the end of their lives, the biggest stars *do* go out with a bang (see Chapters 4 and 9).

But not our Sun. It's a middle-aged, middle-class star and, true to character, does things in moderation. But one day, it *will* run out of the nuclear fuel that powers it. Bereft of fuel, the Sun's core will contract, heating up its outer layers. Our local star will billow out to become a red giant (see Chapter 9) like Arcturus — possibly destroying Earth in the process.

Red giants are notoriously unstable; their distended atmospheres are out of control. Eventually, the star will gently puff its envelope into space. The result is a beautiful ring of gas — a planetary nebula.

These nebulae have nothing to do with planets. They were given their name by William Herschel, discoverer of the planet Uranus (see Chapter 7), who thought that these fuzzy blobs looked like the world he had found.

Planetary nebulae live brief lives — a few thousand years at the most. As it disperses, all that remains of a once-glorious star is its collapsed, shrunken core. This white dwarf star (see Chapter 9) has no energy. All it can do is leak its heat into space, growing dimmer as it cools.

This will be the fate of our Sun: a cold, lonely black dwarf star surrounded by dead planets. But never fear — it won't happen for another seven billion years!

Death of the Sun? The stunningly beautiful Cat's-Eye Nebula represents the future that will befall our Sun billions of years from now.

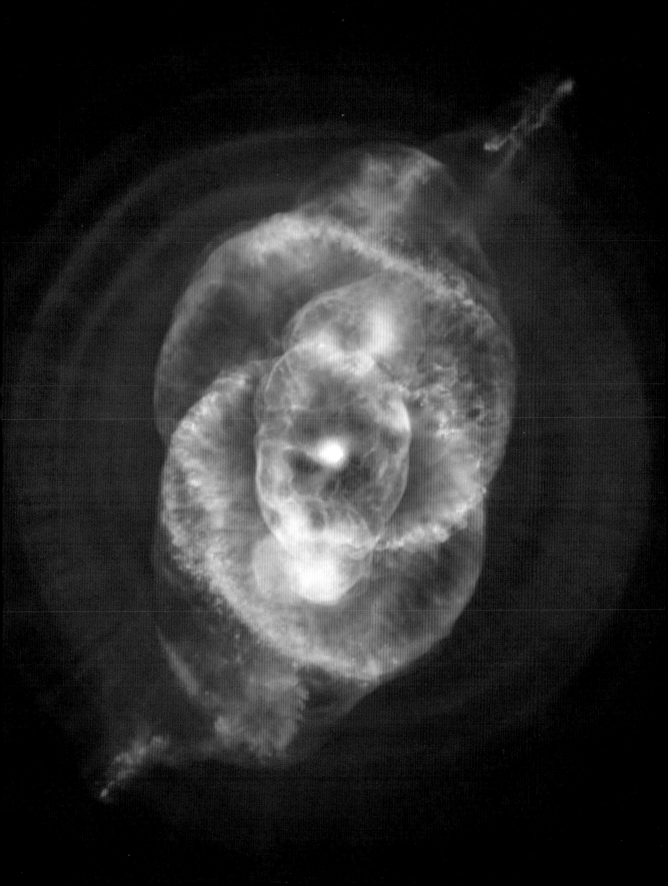

PLANETS 7

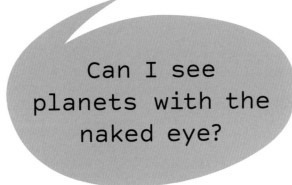

Can I see planets with the naked eye?

Surprisingly, this is the most common question we're asked. Surprising because Venus — the nearest planet to Earth — shines more brightly in the sky than any object except the Sun and Moon. You can't miss it, in the twilight as the morning or evening "star." It can even cast a shadow!

Planets have no intrinsic light of their own; they shine by reflecting sunlight. Venus is swathed in dense clouds of noxious sulfuric acid that reflect the Sun's rays spectacularly.

Although distant, the huge planet Jupiter comes next, tying in brightness with nearby Mars. Both are more luminous than Sirius, the brightest star.

Then comes Mercury, the smallest world in the Solar System. Like Venus, it hugs the Sun in the sky, and — because it's only visible in twilight — you seldom appreciate how bright it is.

Next is Saturn. Smaller and farther away than Jupiter, the ring world is fainter than Sirius.

Now: a test for keen eyesight and dark skies. The planet Uranus lies *just* on the border of naked-eye visibility, although light pollution will wipe it out. Check https://theskylive.com/uranus-info for real-time details of its position.

As for Neptune — it's binoculars or a telescope only.

Four naked-eye planets are captured above the Paranal Observatory in Chile: Venus and Mercury (top); Mars and Jupiter (bottom).

How did the planets get their names?

People have known the five naked-eye planets (see previous question) since prehistory, and the names of these worlds date back into the mists of antiquity. Every culture had its own nomenclature; in the western world, we follow the traditions of the Babylonians, the Greeks and the Romans — going back over two millennia.

The Babylonians and the Greeks set the scene by naming the planets after their gods and goddesses. The Romans followed the system, converting the Greek deities into their own and allotting them planets that suited their temperaments.

Fast-moving Mercury was appropriately named after the fleet-footed messenger of the gods. Dazzling Venus symbolized the goddess of love and beauty (if only they'd known about her searing heat and veils of acid!). Fiery-red Mars took the name of the god of the Romans' major preoccupation, war. Stately and majestic Jupiter took on the mantle of king of the gods. Slow-moving Saturn — the most distant planet they could see — was associated with time and named after the god of agriculture and harvests.

Roll on to 1781 and the first planet to be discovered since antiquity (see Who was the first person to discover a planet?). It was eventually named Uranus — Saturn's father — but only after a slight squabble. It's the only planet to be named after a Greek god.

Neptune came next, and its bluish color evoked the god of the sea.

Which brings us to Pluto. We know it's not officially a planet, but it has a huge fan-following, and the story of its naming is delightful. It was suggested by Venetia Burney, an 11-year-old schoolgirl from Oxford, England. She called it Pluto — after the god of the underworld. How appropriate for a world that many still fondly regard as the farthest planet from the Sun!

It wasn't just a celestial lottery.

Who was the first person to discover a planet?

A German organist and composer living in Bath, England, might seem an unlikely person to discover a planet. But William Herschel was passionate about astronomy and hell-bent on surveying the sky comprehensively. He was also obsessed with building the biggest telescopes of his era. His devoted sister Caroline (an excellent astronomer herself — see page 156) would feed him as he fashioned his enormous mirrors.

One night in 1781, he recorded a greenish blur while surveying the sky. The next night, it had moved. Herschel thought he had discovered a comet, but it was traveling too slowly. Astronomers checked his observations and came to an astonishing conclusion: the amateur William Herschel had discovered a planet.

What to call it — the first world to be found since antiquity? Herschel favored "George," after King George III (the only English monarch known to have had an interest in science). But tradition prevailed, and it became Uranus — the father of Saturn.

The discovery doubled the size of the Solar System. Uranus turned out to be a gas giant like Jupiter and Saturn, but smaller. Nonetheless, it's still four times wider than the Earth and 15 times more massive.

Uranus completes an orbit around the Sun in 84 years. And as it circles the Sun, it spins on its side — probably the result of a massive impact billions of years ago. It's surrounded by a family of 27 moons and a thin set of rings.

Only one space probe, *Voyager 2*, has swung past the planet. It discovered a bland, featureless world. But things are getting more exciting for Uranus. In August 2014, it erupted into giant storms.

Several future space missions to the planet have been planned but none has yet been approved.

Caroline Herschel.

William and sister Caroline Herschel in their drawing room

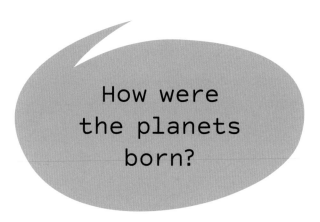

How were the planets born?

Five billion years ago, the planets of the Solar System didn't exist, and neither did the Sun. Instead, there was just a dark cloud of gas and dust floating in space.

A nearby exploding star — a supernova — squeezed the cloud, and it began to collapse. Pulled ever tighter by its own gravity, the cloud contracted into a swirling disc. A dense central hub condensed to become the Sun.

In the disc, microscopic specks of rocky dust stuck together to build up into fluffy clumps, a lot like the dust bunnies under your bed. Out in space, there was no limit to the size of the dust bunnies; they grew up to the size of a city block!

These cosmic bunnies began to collide at high speed, smashing together to make giant space rocks called planetesimals. The inner part of the Solar System was like a stock car race. The colliding planetesimals would often smash each other apart, but sometimes gravity would bind them into a larger world. In the end, a handful of big planets emerged that swept up the remaining planetesimals to become the rocky planets: Mercury, Venus, Earth and Mars. (One final mighty collision with Earth created our Moon — see Chapter 5.)

It was different farther from the Sun. Out beyond the "snow line" (roughly where the asteroids lie now), there was a lot of ice mixed in with the rock. The dust bunnies here were gargantuan snowflakes, and the planetesimals were giant icebergs. That's why we find icy worlds like Pluto, or the big planets Uranus and Neptune, which are vast oceans made from melted ice.

The Solar System's two giants — Jupiter and Saturn — are composed mainly of hydrogen and helium gas. And there are two competing theories to explain how they were born.

According to the first, they were originally a pair of worlds like Uranus and Neptune, but so massive that they could pull in the surrounding gas from the disc to build up their bloated bodies.

The second theory proposes that where Jupiter and Saturn lie in the Solar System, the disc was so dense that two clumps collapsed spontaneously to form two giant gassy worlds, just as the central hub condensed to make the Sun.

What's certain, however, is that a lot of building materials were left over (see next chapter): rocky asteroids, distant icy worlds in the Kuiper belt and the comets that sweep in from the outermost reaches of the Solar System.

This artist rendering captures a very messy business: planetesimals crashed together, in an orgy of destruction and agglomeration, to build today's Solar System.

Is there a planet where it rains diamonds?

Yes, and it's Neptune — the farthest planet from the Sun.

Rather than dropping from the planet's clouds, these gemstones are sinking downward through the giant ocean that makes up most of Neptune — quite appropriate for a planet named after the Roman god of the sea.

The planet's fluid depths are filled mainly with water, but there's some methane mixed in (see page 142). In 2017, scientists at a lab in California used the world's most powerful laser to recreate the immense pressures deep within Neptune, squeezing carbon-rich materials like methane until they condensed into tiny diamonds.

As the diamonds on Neptune rain down toward the planet's core, these sparklers may grow into gems that are millions of carats in weight!

Inside Neptune, the hailstones are pure diamonds.

How can we measure the age of the Earth?

Travel 800 km north of Perth in Western Australia and you come across the Jack Hills, a nondescript, scrubby red ridge. But it holds a great secret. Tiny gems of zircon hidden in its rocks are the oldest surviving material on the Earth, dating back 4,400 million years.

Uranium atoms within the zircon crystals act as a natural clock. They gradually decay into lead, so the amount of uranium that's left tells scientists how old the rock is.

The Jack Hills gems may be the oldest we know of on Earth; but they don't date right back to our planet's birth. These crystals grew in running water — by that time, the planet must have cooled down from its birth pangs and from the incandescent impact that created the Moon.

To find the age of the Earth, scientists turn instead to meteorites. The whole Solar System formed at the same time (see page 136), so researchers hunt for the oldest rocks from space: meteorites that have fallen to Earth, like the lovely specimen in our image.

The birth certificate of the planets turns out to be a type of meteorite called a carbonaceous chondrite. These dark rocks have survived unscathed from the formation of the Solar System. The uranium clock ticking inside the carbonaceous chondrites reveals that they were created 4,567 million years ago. Given the time it would have taken the Earth to assemble from rocks like these, scientists deduce our planet is 4,540 million years old.

Carbonaceous chondrite found in Antarctica: birth certificate of the Solar System

Why are Venus and Earth, twins in size, so different?

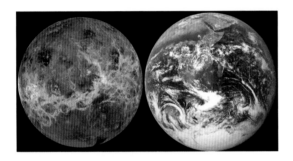

There are two pairs of twin planets in the Solar System: Uranus and Neptune, and Venus and the Earth. The watery giants Uranus and Neptune are fundamentally very similar; Venus and Earth could not be more dissimilar.

Compared to our wet, temperate Earth, Venus is the world from hell. Boasting temperatures of 460°C (860°F) — hotter than an oven — our neighbor planet is the hottest and most poisonous world in the Solar System. Its thick, choking atmosphere is made of carbon dioxide, laced with clouds of sulfuric acid. The pressure at Venus' surface is 90 times that of the Earth.

So if you visited Venus, you'd be simultaneously roasted, corroded, crushed and suffocated!

There are two reasons for the dramatic differences between the two worlds. At a distance of 108.2 million km from the Sun, Venus is intrinsically hotter than the Earth (which lies 149.6 million km away).

But the real culprits are Venus' volcanoes. In the 1990s, NASA's *Magellan* space probe mapped 98 percent of the planet's surface, using a technique akin to aircraft radar to penetrate its clouds. It revealed a staggering number of volcanoes — some of which may be active today. These have created a runaway greenhouse effect that parched Venus dry. So, no possibility of life.

And it's a salutary warning to us as we release increasing amounts of greenhouse gases. Global warming is a reality — not a myth. Tortured Venus is the evidence.

Although future space missions to Venus are on the drawing board, you'll be relieved to know that none involve human crews!

Venus (left) and Earth are both rocky worlds; but our warm and wet planet is the opposite of fiery, inhospitable Venus.

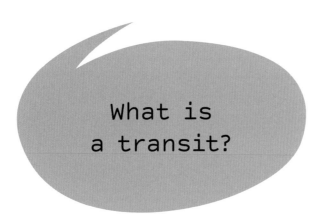

What is a transit?

From our vantage point on Earth we can observe both inner planets, Mercury and Venus, crossing the face of the Sun. These events are called transits.

Transits are mini-eclipses, and the eclipse warnings we gave in Chapter 6 are all relevant here. Observe with care!

Transits take several hours, and it's thrilling to watch these worlds creeping slowly across the Sun's disc. Two transits of Mercury are coming up: on November 11, 2019, and November 13, 2032. Because Mercury is so tiny and relatively far way, its transits are harder to observe.

Venus' transits are much more in-your-face. The planet is larger and closer. The last took place in June 2012 (see image), but, unless you're exceptionally long-lived, you won't catch the next (it's in December 2117).

Before today's technology, transits were a vital scientific tool for measuring the size of our Solar System. In 1769, expeditions from Europe traveled to all the corners of the globe to observe a transit of Venus. The aim was to calculate the distance of the Earth from the Sun — the astronomical unit — a fundamental step in determining the distances of the other planets.

By timing the progress of Venus across the Sun's disc, the teams came up with a remarkably accurate estimate: 153 million km. That's only three million km off from the value we have today (149.6 million km).

Transits are mini-eclipses. Observe with care!

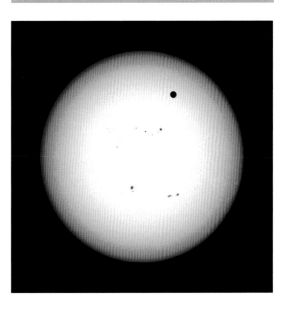

Venus in transit. It's deep black against the Sun's disc, compared to the paler sunspots.

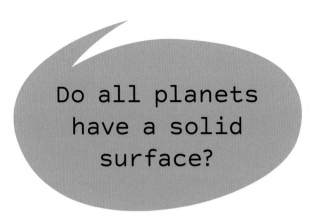

Do all planets have a solid surface?

Years ago, a comet called Shoemaker-Levy 9 crashed into Jupiter. The commissioning editor at an international television channel called us to commission a documentary. "What I want to know," he said, "is what that incredible impact would have looked like from the surface of Jupiter."

We had to tell him that was impossible — Jupiter does not have a solid surface. Fortunately we didn't lose the commission!

In fact, only half the planets have a surface we could stand on: Mercury, Venus, Earth and Mars. Venus and Mars are swathed in a carbon dioxide atmosphere. Earth's air is unique: 78% nitrogen, other trace gases, and 21% oxygen — the product of plant life!

The giants, Jupiter and Saturn, are essentially giant cosmic gas bags. Descend through the mainly hydrogen atmosphere (see next question) and . . . you just keep on going for 50,000 km. As you travel down, the gas becomes thicker and hotter until you're surrounded by liquid, then it becomes a shimmery silver ocean where the hydrogen is so dense it behaves like molten metal. Right at the center, you'll probably hit a rocky core — but it's so hot that it, too, is most likely molten lava with no solid surface.

If you floated down through the atmospheres of Uranus and Neptune, you'd find the dense air merging imperceptibly into a deep ocean of hot water, without even a surface you could dive through. Laced with ammonia and methane, the hot ocean extends down almost to the planet's center. Again, there may be a core of molten rock lurking there.

We haven't gotten very far in exploring the interiors of the giant planets first-hand. In 1995, the *Galileo* probe plunged into Jupiter's atmosphere. It lasted just 58 minutes and traveled only 156 km into the planet before it was destroyed by heat and pressure. The Saturn-orbiting space probe, *Cassini*, was deliberately incinerated in 2017 to prevent it contaminating the planet's moons. Without any protection, it didn't even descend as far as Saturn's cloud-tops before burning up.

The Galileo probe descends through Jupiter's endless atmosphere.

What's the biggest planet?

Not content with being the king of the gods, Jupiter is also the king of the planets. Measuring 147,980 km at its equator, Jupiter could swallow 1,300 Earths and weighs 318 times the mass of our planet.

Jupiter is a gas giant made almost entirely of hydrogen and helium. And, despite its enormous size, it spins faster than any other planet in the Solar System: its day is only 9 hours 55 minutes long. As a result, Jupiter bulges around its equator and is flattened at its poles, making it look like a tangerine when eyeballed through a telescope.

Through a small home telescope, you'll also notice Jupiter's bizarre appearance: it looks like an old-fashioned striped candy. It's girdled by gas clouds stretched into cream and ochre stripes by the planet's rapid spin and furious winds.

You'll also spot Jupiter's four brightest moons, which scamper around the planet from night to night (see page 146).

Jupiter has been visited by a plethora of space probes from Earth — notably the *Pioneers* and the *Voyagers* (which made spectacular flybys), and two orbiters, *Galileo* and *Juno.* The latter is circling Jupiter as we write.

From close-up, we've learned even more about this giant world. It's circled by a faint ring and boasts a ferocious magnetic field, which generates brilliant aurorae and powerful lightning strikes.

Juno is currently on the trail of Jupiter's biggest mystery: what's at the planet's core (see previous question)? We do know that Jupiter's core generates far more energy than it receives from the Sun. It simmers at a temperature of 35,000°C (63,000°F). And this leads to staggering implications for our Solar System.

Had Jupiter been only 75 times more massive, its core would have been hot and dense enough to ignite nuclear fusion reactions. And we'd be seeing two suns in the sky.

Fragments of comet Shoemaker-Levy 9 tear relentlessly into Jupiter.

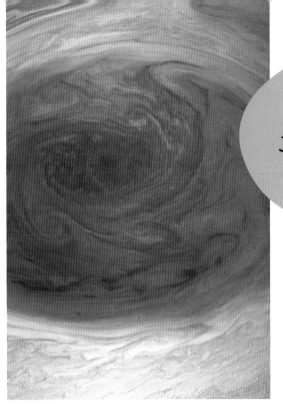

only too well, a hurricane on Earth only loses its power when it moves from the ocean onto land.

And the Great Red Spot probably grew so vast because it swallowed up smaller storms. In 2000, astronomers saw three white spots — small storms — on Jupiter merge to make a single bigger spot. It became a deeper and deeper shade of pink, so it's now dubbed the Little Red Spot.

The Great Red Spot is a gigantic storm in Jupiter's tempestuous atmosphere

That suggests any spot on Jupiter will blush red if it grows large and high enough. The Great Red Spot is so huge that its upper regions tower 8,000 m above the main cloud decks, where they are exposed to the Sun's ultraviolet rays. Most likely, this radiation causes the spot's distinctive color when it reshapes ammonium sulphide in the white clouds into more complex, colored molecules. In other words, the red tint of the famous spot is basically a case of sunburn!

I n 1665, the Italian-French astronomer Giovanni Domenico Cassini saw something remarkable between Jupiter's dark and pale streamers of cloud: "a Spot always adhering to the Southern Belt; its diameter is about the tenth part of Jupiter." This spot was "the most conspicuous and most permanent . . . and appeared to be different in color."

The Great Red Spot is a gigantic storm in Jupiter's tempestuous atmosphere, and Cassini's pioneering observation means it's probably been raging for 350 years. At its largest, a hundred years ago, the Great Red Spot was three times the width of planet Earth, though it's currently shrinking and is now only a little larger than our planet — but that's still one hell of a tempest.

This megastorm has survived so long because the planet has no land masses (see page 142). As people in the southern United States know

Jupiter's Great Red Spot, viewed in close-up by the *Juno* mission.

Which planet has the most moons?

Jupiter is the winner, yet again! Its mighty gravity holds a family of at least 69 moons, forming a miniature solar system. Next comes Saturn, with 62 moons. Uranus is home to 27 moons, and Neptune boasts 14. Okay Pluto fans, we haven't forgotten you; the god of the underworld is circled by five satellites. And the inner planets can't cope with the competition: Mercury and Venus have no moons, Earth has one and Mars has two.

The moons of the Solar System are characters in their own right, not content to live in the shadow of their parent planets.

Take Mars' Phobos. Like its companion, Deimos, it's named after warlike associations — in this case, "fear" and "panic." Phobos is almost certainly a tiny captured asteroid. It circles Mars in a very low orbit and may crash into the Red Planet in some 50 million years' time — blasting out a crater 300 km across.

Jupiter's standout moons are Io, which is peppered with active volcanoes, and Europa, which may harbor life under its icy surface. Saturn's Titan has seas of methane and ethane hiding under its dense atmosphere.

Uranus' tiny moon Miranda is a celestial mess of crumpled craters and cliffs. It was probably smashed apart by a huge impact, but reassembled itself. And Neptune's moon Triton boasts geysers that erupt plumes of nitrogen and dust into space.

What about Pluto? American astronomer Jim Christy discovered its largest moon, Charon, in 1978. It's half the size of Pluto, and Christy named it Charon after his wife, Charlene. Charon also happens to be the rower who ferried souls across the River Styx to the underworld.

When we interviewed Christy for a TV program, we also spoke with Charlene and asked how she felt. Her comment? "Many husbands promise their wives the moon — but my husband got it for me."

Jupiter and its four largest moons, clockwise from top left: Europa, Io, Ganymede and Callisto.

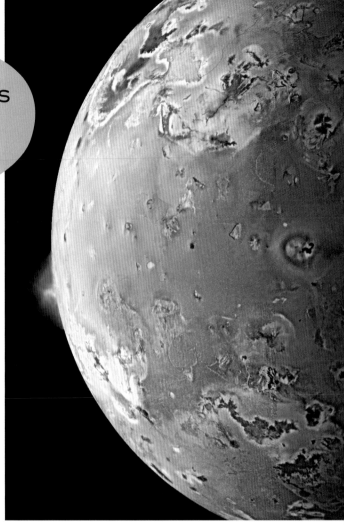

Do other planets have erupting volcanoes?

Earth's volcanoes are iconic. Etna, Vesuvius, Mount St. Helens, Krakatoa — all are signs that our planet is a living, vibrant world, with hot magma welling up from its interior to create these cones of fire. It also boasts some enormous extinct volcanoes: Mauna Kea, on the Big Island of Hawaii, soars to a height of 4,207 m.

What about the other worlds in the Solar System? Venus is a volcanologist's paradise. It is covered in volcanoes that have led to its runaway greenhouse effect.

Our Moon may once have had volcanoes, evidenced by sinuous lava tubes, but they are long gone.

Mars boasts the biggest volcano in the Solar System, Olympus Mons. It's a staggering 26 km high — three times higher than Mount Everest — and, if plonked on the Earth, it would swallow up England. It lives on the Tharsis plateau with other giant volcanoes. Currently it's inactive, but who knows what might happen in the future?

Having no solid surfaces, the outer gas giants aren't volcano territory, but their moons are. Take Jupiter's pizza-lookalike Io. This is the most volcanically active world in the Solar System — it shoots plumes of sulfur dioxide some 300 km into space. The culprit? Pummelling by Jupiter's mighty gravitational field.

Saturn's Enceladus is also a hotbed — or rather, coldbed — of activity. Giant jets of icy water spew into space from under its frozen crust. The particles from these cryovolcanoes have created one of Saturn's fainter rings.

Even colder than Enceladus is Neptune's largest moon, Triton. And like Enceladus, it has volcanoes that erupt plumes of ice.

The Solar System has a lot to keep volcanologists busy.

Spectacular eruption on Jupiter's moon Io

Why does Saturn have rings?

We delight in showing newcomers to astronomy their first sight of Saturn through a telescope. First they gasp. Then they say, "It can't be real — you must have hung a model in front of the lens!"

Yes, Saturn is stunning — not the planet itself, but the amazing set of bright rings which look three-dimensional even through a small telescope. And that's just the beginning. Images from the *Cassini* spacecraft — like the picture above, taken as *Cassini* orbited behind the planet — reveal that the rings we see from Earth (yellow) are surrounded by extensive, much fainter rings (blue.) Uranus, Neptune and Jupiter all have their own ring systems, but they're so faint that you'll need a space probe to find them.

Though Saturn's main rings are wide enough to stretch almost from the Earth to the Moon, they are very thin — less than 100 m thick. If you wanted to make a scale model of Saturn's rings out of paper, they'd need to be wider than the length of a football field.

Astronomers know that Saturn's rings aren't solid. They're made of a gazillion chunks of ice, ranging in size from snowballs to icebergs, racing around the planet like minuscule moons. But nobody is sure why Saturn has this unique attraction. Perhaps they were simply rubble left over from the planet's construction (see page 136).

Or perhaps they're much younger — though "young" on the cosmic scale means around 100 million years old. The *Cassini* space probe (which orbited Saturn from 2004 to 2017) found that the ice in the rings seems remarkably fresh — untarnished by space dust that fills the Solar System.

If the early dinosaurs had had a telescope, they would have seen Saturn merely as a planetary globe. A hundred million years ago, an icy moon got smashed up — either when it was hit by a wayward comet, or just because it strayed too close to Saturn — and its remains spread out to create the most glorious sight in the Solar System.

Even Saturn's faintest rings shine when they're backlit by the Sun.

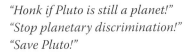

Is Pluto a planet?

"Honk if Pluto is still a planet!"
"Stop planetary discrimination!"
"Save Pluto!"

What was all that about? Well, in 2006, the International Astronomical Union voted to demote Pluto from its planet status.

Back in 1930, a young American called Clyde Tombaugh discovered a new world beyond Neptune, at the frontier of the Solar System, and everyone called it the ninth planet of the Solar System.

But Pluto came down with a bump 76 years later. Astronomers had found that Pluto shares its territory with a large host of minor worlds in the Kuiper belt (the icy equivalent of the asteroid belt between Mars and Jupiter). And in 2005, astronomer Mike Brown in California discovered a celestial body in the Kuiper belt that is a twin to Pluto. He named it Eris (appropriately, meaning "discord").

So should Eris take on the rank of a planet? Or were these little runts of the Solar System just members of a swarm? The International Astronomical Union went for the second option, and both Pluto and Eris have officially become dwarf planets!

Dwarf or not, Pluto, which is about two-thirds the size of our Moon, is a fascinating object. In July 2015, NASA's *New Horizons* space probe flew past the tiny world. Amazingly, for an icy denizen at the edge of our Solar System, it seems to be geologically active. Its surface is smooth and appears to be renewing itself from within. Pluto also boasts an atmosphere of nitrogen, which is streaming away into space.

It is accompanied by five moons. The largest, Charon (discovered in 1978), is distorted by a spectacular gash.

New Horizons has headed out into the Kuiper belt to explore another icy dwarf, Ultima Thule. Who knows what kind of weird and wonderful world it will find?

The white heart discovered by *New Horizons* is named Tombaugh Regio, after Pluto's discoverer.

Could there be unknown big planets at the edge of the Solar System?

There's something lurking out beyond the known planets — and it's BIG . . .

Since 1992, astronomers have found over 2,000 icy worlds orbiting the Sun farther out than Neptune. Since 2006, Pluto has been classified as one of these Kuiper belt objects. But they're all tiny compared to the other planets. Astronomers reckon that if you put together everything in the Kuiper belt, you'd make an object far smaller than the Earth.

There are some odd objects out in the Kuiper belt — small worlds that follow very elongated orbits. Even stranger, many of these orbits seem to line up. In 2016, two astronomers at Caltech — Konstantin Batygin and Mike Brown — suggested that the gravity of a distant planet was marshaling these Kuiper belt objects.

Rather cheekily, they called it Planet Nine. Pluto had been regarded as the ninth planet from its discovery in 1930 until Mike Brown led the campaign to demote Pluto (see previous question) in 2006. He even uses the Twitter handle @plutokiller.

Batygin and Brown's proposed Planet Nine certainly puts diminutive Pluto in the shade.

They calculate the new planet is about 5,000 times heavier, making it 10 times more massive than the Earth. Why has no one spotted such a whopper? Because, they say, it lies way out in space, 25 times farther than Neptune.

In 2017, Kat Volk and Renu Malhotra of the University of Arizona stirred the mix more by announcing a possible Planet Ten. This suspected world would lie much closer than Planet Nine, about twice Neptune's distance from the Sun, and it would be about the same size as Mars. Volk and Malhotra calculated that the Solar System needs Planet Ten to explain why the orbits of Kuiper belt objects are tipped up.

The evidence for Planet Ten is still weak, but many astronomers around the world have picked up on the much stronger scent of the distant giant, Planet Nine. As we write these words in late 2017, many astronomers are now hot on its trail, scrutinizing every corner of the constellations Orion and Cetus, where it's most likely to lurk. By the time you read these words, the Solar System may have welcomed another major world to its family of planets.

Looking back at the Sun from "Planet Nine"

COMETS, ASTEROIDS AND METEORITES

8

What's a comet?

"When beggars die, there are no comets seen; the heavens themselves blaze forth the death of princes."

Calpurnia spoke these words to her husband, Caesar — in Shakespeare's Julius Caesar — sensing a portent of his death on the morning of the Ides of March (the 15th of the month).

Of all astronomical objects, comets are those most associated with doom and destruction. They appear as ghostly interlopers in our sky. With their luminous tails stretching in their wake, they're unlike anything else in the heavens. An anonymous 1910 poem about Halley's comet puts these celestial beasts in their place: "You ain't made of nuffin. An' mostly gassified."

This more or less sums comets up.

Comets are something made out of nothing. Measuring only a few kilometers across, these survivors from our planets' birth live in a vast cloud — the Oort-Öpik Cloud — that surrounds our Solar System and may stretch halfway to the nearest star.

How many comets are there? Estimates run into billions and trillions. What's more certain is that the Sun has little control over these outer realms of the Solar System. The pull of a passing star can knock a comet off its perch, making it plummet inward toward the Sun — and that's when the fun starts.

Astronomers call comets "dirty snowballs." An apt description; they're made of ice and rock. As a comet plunges sunward, a spectacular transformation takes place. The Sun's heat boils away the ices of this cosmic runt, and its gases billow out as a glowing coma, which can be as large as the Sun itself.

And the Sun's energy drives the gases back into two spectacular tails, which can be millions of kilometers long. The blue gas tail — energized by the solar wind of electrically charged particles — points directly away from the Sun. The curved, yellowish dust tail is pressured by the Sun's light.

After the comet rounds the Sun (unless, in some unfortunate cases, it collides with it), it's downhill all the way. It may head back to its home in the Oort-Öpik Cloud, or remain trapped in the Solar System by the gravity of the planets.

Comet Hale-Bopp — the Great Comet of 1997

How fast do comets rush across the sky?

We always enjoy the chance to be interviewed when there's a new comet around. After all, they are fairly rare visitors to our skies, and people love to view them. After the initial chat, our hearts fall when we're asked live on air: "So the comet's visible tonight; what time do we have to be out to see it flash past?"

It's just not like that. A comet isn't a shooting star, whizzing past in a brief instant of glory. A comet hangs in the sky for weeks or months — or even over a year, in the case of the brilliant comet Hale-Bopp of 1997. Our image shows the wonderful many-tailed comet McNaught, which enlivened the Southern Hemisphere skies early in 2007 and hung in there for several weeks.

OK — if you Google the speed of a comet, you'd find some impressive statistics. For instance, when Hale-Bopp was closest to the Sun, the comet was hurtling through space at 156,000 km/h. But Hale-Bopp was also a long way from us — and the farther away something is, the slower it seems to move across the sky. Think of watching a plane: stand next to the runway, and it rushes past you in an instant; but watch an aircraft high in the sky, and it seems to dawdle.

So a comet may be rushing through space, but from our distance we have a leisurely view of its magnificent passage through the heavens.

Comet McNaught and its spectacular tail, seen in 2007

How are comets named?

Want to become immortal? Easy: discover a comet.

Traditionally, if you spot a new comet, it's named after you. So we all call the magnificent apparition of 1997 comet Hale-Bopp, as it was found independently by American astronomers Alan Hale and Thomas Bopp. But comets also have grown-up names, allocated by the International Astronomical Union, which keeps celestial events in meticulous order. Hale-Bopp is officially C/1995 O1, because it was a comet (C) discovered in 1995; the O1 gives its order of discovery within that year.

The first person to discover a comet was German astronomer Gottfried Kirch, in 1680. The first discovery by a woman was by William Herschel's sister Caroline, in 1786 — although it's claimed that Kirch's wife, Maria, also discovered a comet, but its finding was credited to her husband.

Caroline would go on to find eight comets in total — and was awarded a salary by King George III for her efforts. In essence, she was the first female professional astronomer.

Comets are allowed to have three names. One classic is comet IRAS-Araki-Alcock, discovered by a satellite (IRAS, the Infrared Astronomy Satellite) and two school teachers: Genichi Araki (Japan) and George Alcock (UK). In 1983, it approached Earth closer than any comet for 200 years: just 4.7 million km away.

Our other favorite is the tongue-twisting comet Churyumov-Gerasimenko (known to its aficionados as comet C-G), which was jointly discovered by two Russian astronomers. In 2014, Europe's *Rosetta* spacecraft sent the probe Philae to explore the comet's surface and get a better understanding of the birth of the Solar System.

How does it feel to discover a comet and follow its unpredictable behavior around the Solar System? Canadian amateur astronomer David Levy, namesake of 22 comets, has said: "Comets are like cats. They have tails, and they do precisely what they want!"

Eighteenth century satirical cartoon of Caroline Herschel discovering a comet. We don't think prim Miss Herschel would have been impressed!

What's the brightest comet ever seen?

I f you were lucky enough to see comet Hale-Bopp in 1997, or comet McNaught 10 years later, you know that a comet can appear brighter than any star. But these recent cosmic visitors don't hold a candle to some comets that amazed our forebears.

The brightest comet of the twentieth century was Ikeya-Seki, which swept just a million kilometers from the Sun's blazing surface in 1965. People could see it in broad daylight just by blocking out the Sun's bright disc with their hands.

Our striking image shows a similar comet that appeared in the previous century, lying to the upper right of the Sun above Table Mountain. Eyewitnesses described the Great Comet of 1843 as "an elongated white cloud" and a "short, dagger-like object" close to the Sun in the daytime sky. We know this portrayal is as accurate as you can get, because it was painted by an astronomer — Charles Piazzi Smyth, a British scientist who was working in South Africa at the time (in case you think that's a slightly un-English middle name, his godfather was Giuseppe Piazzi, who discovered the first asteroid — see page 165).

But the greatest comet ever recorded was the Great September Comet of 1882. As it skimmed only 400,000 km above the solar surface, the comet was far brighter than the Full Moon. In fact, astronomers could view it through a dense, dark filter — like the eclipse goggles we use to observe the Sun — and saw that "the silvery light of the comet presented a striking contrast to the reddish-yellow of the Sun."

Her Majesty's astronomer in Cape Town, David Gill, was so carried away by the sight of this dazzling comet in the dawn sky that his sober scientific report breaks into a poetic eulogy: "An ill-defined mass of golden glory rose with a beauty I cannot describe."

The daytime comet of 1843, to the upper right of the Sun, painted by eyewitness Charles Piazzi Smyth.

What's special about Halley's comet?

"Of all the meteors in the sky, / There's none like Comet Halley. / We see it with the naked eye, / And periodically."

Even professors of astronomy at Oxford — in this case the nineteenth-century Herbert Hall Turner, who penned these lines — get things wrong. He got comets and meteors mixed up, as most people do.

He did get one thing right: comet Halley is iconic because it returns regularly to our celestial neighborhood. It's trapped in the Solar System, and swings by the Earth every 76 years — an uncanny resemblance to the human lifetime. So we associate our own encounters with the comet as a once-in-a-lifetime experience.

Despite its name, the most famous comet of all *wasn't* discovered by Edmond Halley. He first saw the comet on his honeymoon in 1682. His friend — the reclusive and introverted Isaac Newton — had just come up with his theory of gravity, which predicted how celestial bodies moved. Armed with Newton's math, Halley went off to calculate the orbit of his object — and it took six weeks to complete the sums! Halley realized that comets seen in 1531 and 1607 were actually the same object returning regularly to the Sun. Boldly, Halley predicted it would return in 1758.

It did. Halley's fame was assured, and the comet was rightly named after him.

Comet Halley has been making regular appearances in our skies since at least 240 BC (next visit in 2061). Many of its visits have been documented. It is often linked to the Star of Bethlehem (see Chapter 2), but the dates don't add up.

One in the eye for Harold: Halley's comet as a portent for England's king in the Bayeux Tapestry.

We're on much firmer ground with 1066. The comet hovered in the sky as Duke William of Normandy prepared to invade England. King Harold's advisors saw it as an omen. At the Battle of Hastings, they were proven right: Harold was reputedly killed by an arrow-shot in the eye.

The comet's last appearance, in 1985–86, was very disappointing — particularly from the Northern Hemisphere. But south of the equator it put on a better show. We were lucky enough to see it from a high-flying supersonic Concorde plane on a lecturing trip to New Zealand. Europe's space probe *Giotto* made its own trip to the comet. It audaciously flew *through* the comet, discovering the nature of its heart: a potato-shaped body just 15 km long.

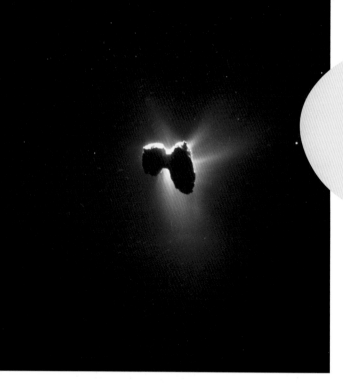

Are comets all bad news?

Comets have always had bad press. And it's easy to understand why. A lot has to do with their appearance: they can look ghostly and threatening, hanging motionless in the dark sky like a dagger about to strike.

And strike they do. Sixty-six million years ago, a comet (or possibly an asteroid) hit the Yucatan Peninsula on the coast of Mexico. It blasted out the biggest impact crater in the world. The impact created a burning sky that enveloped the Earth, setting fire to its forests.

Over 75 percent of all species living on Earth were wiped out — roasted to death. And that included the mighty dinosaurs that had ruled our planet for more than 100 million years.

And if you want to see the havoc that comets can wreak, look back no further than July 1994. That's when a comet collided with mighty Jupiter, bringing real-time drama to our screens.

Comet Shoemaker-Levy 9 had been discovered by the dedicated comet hunters Gene and Carolyn Shoemaker and their colleague David Levy the year before. It was in orbit around Jupiter, and it quickly became evident that it was on a collision course.

And *what* a collision! We Earthlings have never seen the likes of it before. The gravity of the giant planet tore the unfortunate comet apart. We'll never forget the images of fragments from the doomed comet smashing into the gas giant.

Traveling at speeds of over 200,000 km/h, 21 pieces of Shoemaker-Levy 9 created giant black scars on Jupiter that lasted for months.

Despite this trail of destruction, astronomers are now seeing a kinder side to these cosmic vagabonds. Impacts in the distant past may have been largely beneficial. Fresh comets contain a lot of water, and those that have hit the Earth could have contributed in a major way to our planet's oceans.

So next time you see a comet, gaze upon its beauty and contemplate that these dirty snowballs might have brought life-giving water to our world.

Duck-shaped nucleus of comet Churyumov-Gerasimenko, imaged by the Rosetta probe in 2016

What is a shooting star?

When I, Heather, was 7, I had a burning ambition to be a pilot, like my dad. I'd spend nights gazing up at the planes stacking over nearby Heathrow Airport.

One night, I saw something amazing: a green meteor. I rushed into my parents' living room to tell them (it was well after my bedtime), but they dismissed my discovery. "There's no such thing as a green meteor, dear."

On the front cover of the *Daily Express* the next day was the headline: "Green Shooting Star Seen over West London." That was it. I wasn't going to be a pilot; I was destined to become an astrophysicist. And that's where it all ended up . . .

But as for the meteor? Amazingly, although a "shooting star" does indeed look like one of the stars falling from the sky, it's nothing but a tiny speck of dust — rather like a granule of instant coffee in size and texture. It looks so brilliant because the meteor hits the Earth's atmosphere at an incredible speed — up to 250,000 km/h — so it burns up in a blaze of glory.

A few of these celestial fireworks have gaudy colors. Why was mine green, for instance? It was probably caused by the little critter tearing up oxygen atoms in the Earth's atmosphere, making them glow green.

You can see a few meteors every night of the year — they're debris strewn from comets as they traipse around the Sun. But there are special times to observe shooting stars: August and December. That's when our planet intercepts regular streams of particles from visiting comets. Meteor showers are named after the constellations from which they appear to fall. So look for the Perseids in summer, and the Geminids in winter.

There's just one question that we scientists still can't answer: If you wish on a shooting star, will your dreams come true?

Spectacular green meteor captured over the town of Mettupalayam in India

Where are the asteroids?

There's something odd about the Solar System. Close to the Sun we've got four rocky planets, including Earth, in tight little orbits. In the outer parts, you'll find the four gaseous planets lazily pursuing their large orbits, and then small icy objects like Pluto and the comets. But there's a large gap between Mars, the outermost of the rocky worlds, and Jupiter, the first of the gas giants.

Look closer though, and the seemingly empty gap is filled with millions of rocky chunks: the asteroids. The biggest, Ceres, is less than a thousand kilometers across — about the size of Texas — and the smallest no bigger than a house.

Despite the vast number of asteroids, they are so small that if you lumped them all together you'd make a world far smaller than our Moon.

The asteroid belt between Mars and Jupiter is home to most of the asteroids, but some of them like to stray. The Trojan asteroids share the same orbit as Jupiter, either so far behind or so far ahead of the giant planet that they are in no danger of collision.

We need to be more worried about asteroids that have escaped the asteroid belt and are headed toward the Sun, because they could be on a collision course with Earth (see see page 174).

Speeding rocks on the interplanetary beltway

How were asteroids made?

When they were first discovered, astronomers thought the asteroids were shrapnel from a world that had exploded. But now we know asteroids are the remains of The Planet That Never Was . . .

Back in its early days, the inner part of the Solar System was filled with giant space rocks called planetesimals (see page 136). Most of them eventually stuck together to make up the planets Mercury, Venus, Earth and Mars. But beyond the orbit of Mars, things were different. The neighboring planet, Jupiter, was constantly meddling, its powerful gravity stirring the planetesimals and preventing them from gathering into a single planet.

In fact, Jupiter didn't just disrupt the birth of a planet in the asteroid belt — it flung most of the planetesimals away into space. A thousand times more material was thrown away than survives as today's asteroids.

At first, some of the asteroids were quite big — as Ceres is today — with molten rock inside and a center so hot that iron flowed inward to make a core of liquid metal.

But the asteroid belt is a violent place, with asteroids colliding throughout its history. Many of the biggest denizens were smashed apart. Their surface layers became dark asteroids, with a composition that's little changed since the birth of the Solar System. The shattered interior ended up as brighter rocky asteroids, while blobs from the core have created asteroids made of metal. Cosmic mining companies plan to extract these metals to sell back on Earth.

Asteroids collide to create a family of smaller space rocks.

Engraved by C. Turner.

Who discovers and names new asteroids?

On the night of January 1, 1801, the Italian astronomer Giuseppe Piazzi was mapping all the stars in the constellation of Taurus, the bull, when he spotted a star that shouldn't have been there. The next night, it had moved. It was clearly a small world orbiting the Sun beyond Mars.

Ironically, a letter was in the mail to this astronomer on the island of Sicily from a consortium of astronomers based in Germany who were banding together to look for a planet in what was then an empty gap between Mars and Jupiter. They called themselves the Celestial Police, and Piazzi had unknowingly apprehended the culprit. Still, it clearly wasn't a major planet, and the celestial officers wondered if it was part of a gang. Sure enough, they soon had three other small bodies in their sights. Because these worlds were so small that they looked just like points of light, astronomers confusingly called them *asteroids,* meaning "star-like."

Since then, more and more powerful telescopes have revealed asteroids by the hundreds, then the thousands, and now the hundreds of thousands. They often give themselves away by leaving annoying tracks across long-exposure photographs of the sky, and irate astronomers call them "vermin of the skies."

But how do astronomers name all these tiny pests? For a start, as every asteroid is found, it's given a number. And then the discoverer has the right to give it a name. Piazzi called the first asteroid Ceres, after the patron goddess of Sicily, and it was followed by Pallas, Juno and Vesta.

There was a limit to the female deities inhabiting Mount Olympus, however, and fresh generations of asteroid hunters had other great figures to immortalize: Einstein, Beethoven and Gagarin, for instance — not to mention Lennon, McCartney, Freddie Mercury and Bilbo (the hobbit).

And things started to get out of hand. It was bad enough when French astronomer Auguste Charlois named five asteroids after his mistresses (he was later murdered by his brother-in-law). But then Mr. Spock appeared in the asteroid belt — not the *Star Trek* character, but the discoverer's cat. The International Astronomical Union now has to vet all the proposed names. So we were pretty stunned when they honored the two of us with asteroids: number 3795 is now "Nigel," and asteroid 3922 is "Heather."

Asteroid pioneer Giuseppe Piazzi was the first to name a space rock.

How dangerous is it to fly through the Asteroid Belt?

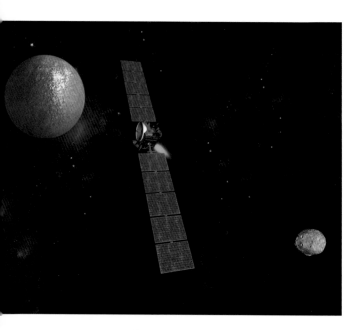

Surely NASA must have been joking? It was 1972, and the space agency launched an unmanned spacecraft toward Jupiter — straight through the asteroid belt. *Pioneer 10* seemed to be heading toward certain destruction as it ran the gauntlet of millions of rocky bullets traveling at 65,000 km/h.

Sister ship *Pioneer 11* was held in check as scientists monitored the fate of *Pioneer 10.* And after seven months of tension, the *New York Times* headline said it all: "*Pioneer 10* Pierces Asteroid Belt Safely."

To be fair, NASA's scientists weren't as worried as the media. Space is a huge place, and there are yawning gaps between the asteroids. *Pioneer 11* was dispatched safely to Saturn. To date, eight other missions have safely followed. In fact, when NASA wanted the *Galileo* probe to image a couple of asteroids on its way to Jupiter, they had to carefully aim it toward Gaspra and Ida, or *Galileo* wouldn't have seen any asteroids at all!

Half a dozen other spacecraft have completed similar feats of celestial navigation to view asteroids at close quarters. In 2011, the *Dawn* spacecraft not only reached Vesta, but slipped into orbit around it, and a year later departed to orbit the largest asteroid, Ceres (which the International Astronomical Union confusingly calls a "dwarf planet"). Astronomers now see the asteroid belt not as an interplanetary minefield, but as a scientific playground packed with clues to the origin of the Solar System.

Dawn investigated Vesta (right) and Ceres — but they were never this close together!

Can I buy a meteorite?

Yes — you can acquire your own mini space-rock. Just Google "meteorite"! Many meteorites reside in museums and labs, where scientists can analyze them for clues to the origin of the Solar System. But people who discover meteorites can, and do, sell them on the open market.

Top prices are reserved for stones that have been blasted off the Moon or Mars.

Size isn't all that determines price: rarity, notoriety and origin all push up the value. You can buy a small piece of a stony or iron meteorite — picked up by the bucketload in the Sahara — for as little as $20. If you want an uncommon carbonaceous chondrite containing organic molecules, you may be looking at 10 times the price.

And you'll pay more for even a common iron or stone if it has a story to tell, such as an iron meteorite from the famous Meteor Crater in Arizona, or a stony fragment (see picture) of the huge meteorite that lit up the sky above the Russian city of Chelyabinsk in 2013, smashing thousands of windows but miraculously not killing anyone.

Top prices in the meteorite market are reserved for stones that have been blasted off the Moon or Mars. If you want more than a speck, you're looking at hundreds of dollars and up. In 2012, a 2 kg chunk of Moon rock fetched $330,000. But that's outclassed by a piece of Martian meteorite weighing a mere 188 g that was sold for a staggering $450,000 — on eBay!

For sale: a fragment of meteorite nestled in a presentation case, with details of where and when it fell to Earth

Where do meteorites come from?

People get mixed up about meteors and meteorites. "Meteorites" sound like little meteors, but actually they're the big ones. Meteors are tiny grains of dust the size of coffee granules shed by comets as they move around the Solar System. When they hit the Earth's atmosphere, they burn up, and we get to see a shooting star.

Meteorites are fragments of minor planets that have suffered collisions.

Meteorites are made of sterner stuff. They, too, plunge into our atmosphere, but they're big enough to survive the journey to the ground. Some meteorites are as small as pebbles. Then there's a huge range of increasing girth up to the world's record-holder, the 60-ton Hoba West meteorite in Namibia. This huge chunk of iron was discovered around 1920 by a farmer plowing a field with his ox. It's now a tourist attraction.

Meteorites mainly come from the asteroid belt — they're fragments of minor planets that have suffered collisions. Our picture shows two thin slices of a meteorite found in Antarctica. There's strong evidence it was once part of Vesta, the brightest of the asteroids (it's *just* visible to the unaided eye). When NASA's spacecraft *Dawn* orbited and explored Vesta in 2011–12, it discovered two enormous craters on the asteroid's surface. The blasts that created the craters splattered rocks throughout the inner Solar System, and some fell on the Earth as meteorites. Vesta's surface is made of crystallized magma, and the multitude of minerals encased in this Antarctic meteorite match the composition of Vesta exactly.

Most meteorites, like those from Vesta, are stony. Six percent are made of iron, chips from the once-molten metallic cores of asteroids that have been smashed apart. But the ultimate meteorites cherished by scientists are carbonaceous chondrites. These are carbon-rich rocks from an asteroid's surface that haven't changed for billions of years. They're the time machine that gives us the deepest insights into the history of the Solar System.

The asteroid belt doesn't have a monopoly on meteorites. We have meteorites from the Moon, and Martian meteorites too! ALH84001, discov-

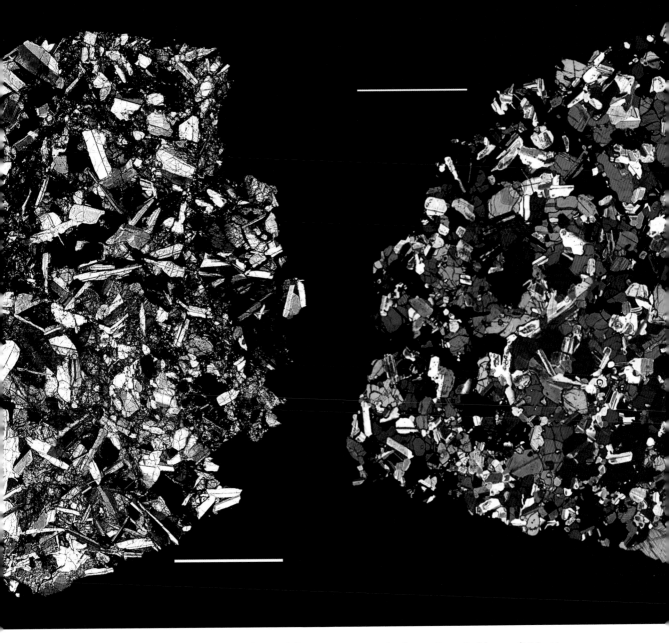

ered in Antarctica in 1984, was blasted out of the Red Planet by a massive cosmic impact. Twelve years later, NASA scientists announced that it contained fossilized microbes from Mars.

Alas, it was not the case. After more analysis, scientists have concluded that the "wormlike" structures in the rock probably aren't Martian fossils after all, but fragments that have broken off crystals within the meteorite. But don't give up on life on Mars (see Chapter 14).

Slices of meteorites from asteroid Vesta, viewed through a microscope

How do scientists find meteorites?

The biggest meteorite ever seen to fall to Earth blazed through the atmosphere in 1947 over the Sikhote-Alin Mountains in the eastern part of the then Soviet Union. "A piece of the Sun has fallen off!" screamed a schoolgirl. Just a year and a half after Hiroshima, others thought it was an American atomic bomb.

A local artist, P.I. Medvedev, was just setting up his easel to paint a landscape when the fireball flashed across the sky. His picture of the incandescent fireball appeared 10 years later on a postage stamp.

From these eyewitness accounts, scientists tracked where the Sikhote-Alin monster must have landed. Because it shattered passing through the atmosphere, the 100-ton meteorite didn't punch out a big crater. Instead, wreckage was spread over a large area (the "strewn field"). Searchers have picked up over 9,000 fragments of this iron meteorite.

A chunk of space debris that's seen to arrive on Earth like this is known as a "meteorite fall," and they are especially important to scientists. From the witnesses' reports, you can work out the meteorite's path in 3-D and deduce where it came from before it hit our planet. And the fragments are delivered absolutely fresh from space, without being eroded by Earth's rain, snow and microbes.

Of course, there isn't always someone around to catch a "falling star." Many meteorites have fallen unrecorded and unseen. When scientists discover one of these space rocks, it's called a "meteorite find."

But how do you track down a meteorite among all the other rocks littering the Earth's surface? Many people find odd stones that they think may be space rocks, but only one in a thousand sent to museums for identification turns out to be the real McCoy.

One sure signpost is a hole in the ground blasted out by an impact from space. Collectors have found thousands of iron meteorites littering the ground around Meteor Crater in Arizona (see page 172).

Or you can hunt for dark-colored meteorites that have fallen on a bright region of the Earth's surface. If it's a desert, then the space rocks may have accumulated here over thousands of years

An eyewitness view of the great meteorite fall of 1947, on a stamp issued 10 years later.

without weathering away. The vast Sahara Desert is a fertile hunting ground for meteorite falls, as is the bright white limestone of the Nullarbor Plain in Australia. But by far the largest number — over 22,000 meteorites — have been discovered in the brilliant ice sheets of Antarctica.

To find a fallen meteorite, however, you sometimes just need luck. Texas Christian University in Fort Worth displays the biggest meteorite ever found in that state. It was discovered by local rancher Frank Hommel in 2016 on his way to a watering hole with his horse Samson. The stallion stopped at an odd rock poking out of the ground and refused to move on. When Hommel and his wife dug up the stone, experts found it was a meteorite weighing a third of a ton.

Members of the Antarctic Search for Meteorites program collect a meteorite that fell thousands of years ago.

Where can I visit a meteorite crater?

This enormous hole in the ground was blasted out by a small iron asteroid that smashed into the Arizona desert around 50,000 years ago. It's a spectacular sight at sunset. We remember flying right over it en route to LA and seeing the bowl deep in shadow, with the rim glowing golden in the light of the setting Sun.

It's the best-preserved meteorite crater in the world, and one of the easiest to visit. Be prepared for its scale; it's over a kilometer across and nearly 200 m deep. The crater is big enough to swallow 50 New York City blocks. And at its center, only the tallest skyscrapers would poke up above ground level.

The Earth boasts around 190 impact craters, as well as tantalizing hints of bigger blast structures that remain to be confirmed. North America lays claim to the most craters (60), including two sensational craters in Canada: Sudbury in Ontario, and Manicouagan in Quebec. Sudbury, gouged out 1,849 million years ago, is 250 km across. It's the second-largest impact structure on Earth, and one of the oldest.

Manicouagan (100 km in diameter) was blasted out 215 million years ago by a meteorite 5 km across. The impact created an annular lake around the crash site, which is used for generating hydroelectric power. The crater's so huge that you can see it easily from space.

If you're visiting Australia, head for Henbury Crater in the country's Northern Territory. The site consists of 13–14 craters created when a meteorite broke up over the region around 4,000 years ago. The largest crater is 200 m across.

Astronomically speaking, this is a very recent impact. The native people were living in the region at the time and have incorporated the craters into their folklore right up to this day.

Arizona's spectacular Meteor Crater. It measures 1.2 km across.

Has anyone been killed by a meteorite?

The short answer is "no" — though a Venezuelan cow was struck dead by a 50 kg meteorite in 1972. The cow was turned into premium steak, while fragments of the "killer meteorite" sell at a premium price.

Back in November 1954, Ann Hodges was luckier. While she was having an afternoon nap on her couch, at home in Sylacauga, Alabama, people outside were treated to the sight of "a bright reddish light like a Roman candle trailing smoke" racing through the skies. The first Ann knew was when a grapefruit-sized meteorite smashed through the roof, bounced off her large radio set and hit her on the side. She suffered a massive bruise, but fortunately no internal damage.

In 1992, student Michelle Knapp was safer indoors than out. Hearing a deafening sound outside her home in Peekskill, New York, "like a three-car crash," she rushed outdoors to find a huge dent in her pride and joy, a red Chevy Malibu she'd just bought from her grandmother for $400. A massive stone smelling of sulfur had sliced through the back of the car, narrowly missing the fuel tank.

Thousands of people in the eastern United States had seen this cosmic vandal streaking across the sky from space, and many had videoed it as the meteorite lit up their Friday night high-school football games. Her insurance company refused to pay up, saying the fender bender had been "an act of God." But Knapp had the last laugh, selling the meteorite — and her uniquely reshaped car — for a fortune!

Rear-ended by a space rock, in Peekskill, New York

Are we going to be wiped out by an asteroid or comet?

The residents of Chelyabinsk in Russia had a near-death experience in 2013 (see page 167). A 20 m asteroid narrowly missed the city, causing extensive damage but no deaths.

'Twas not the same for the dinosaurs, and for most of the other species of their era. Sixty-six million years ago they were wiped out by a comet or meteorite (see page 160). It was the last major mass extinction of life on Earth, but by no means the only one.

> ## Sixty-six million years ago dinosaurs were wiped out by a comet or meteorite.

Our planet's most recent brush with disaster took place on June 30, 1908. A body estimated to be 60–190 m across exploded over a remote region of Siberia, near the Tunguska River. There were loud bangs and a massive shockwave that knocked people off their feet. The driver of the Trans-Siberian Express thought that his train had been derailed! Eyewitnesses reported a bluish-white object in the sky, which disintegrated. It didn't appear to hit the ground.

An expedition reached the region in 1927. Some 2,000 square km of forest had been flattened, felling 80 million trees. There were no human fatalities, but scientists discovered the charred bodies of hundreds of reindeer.

Stitching together the evidence, it seems that the object exploded 5–10 km up, generating the energy of 1,000 atomic bombs. Almost certainly it was a comet. Unless they're large, these fragile objects can't get all the way down through Earth's atmosphere.

When are we due for the next impact? Astronomers have been mapping the locations of NEOs — Near-Earth Objects that have the potential to hit Earth — for some years. So far, 16,000 asteroids and 100 comets are on the danger list.

In 1998, NASA was charged with Project Spaceguard: to seek out objects more than a kilometer across that could cause global devastation.

Scientists have also been combing the records to ascertain the frequency of previous mass extinctions. The results are controversial, but

generally range from 20 to 30 million years.

So if a rogue object is heading our way, what do we do? Sending a space probe to intercept it and deflect it won't work; it takes around four years to plan and launch a mission. The best we can hope for is to plot its course and evacuate the impact area—and cross our fingers.

Cosmic oblivion for life on Earth is still in the cards.

STARS 9

What are the Pillars of Creation?

It's one of the "100 Most Influential Images of All Time," according to *Time* magazine — like Neil Armstrong's shot of Buzz Aldrin on the Moon. The Pillars of Creation is the most astonishing picture from the Hubble Space Telescope, and this sensational cosmic portrait rocked the world when it was first published in 1995. For so many people, it opened their eyes to the staggering beauty of the universe. Hubble revisited the Pillars in 2014, capturing this even more detailed image.

Located in the Eagle Nebula, the Pillars are part of a complex of starbirth in the constellation of Serpens. Lying some 6,500 light years away, the Eagle Nebula reveals the intricacies involved in the process of making stars.

The Pillars are still active in star formation. Each is 5 light years high. Astronomers appropriately call them "elephant trunk structures," and they are created by violent winds from young, hot stars.

The once-dark cloud from which the Pillars arose is lit by the searing ultraviolet light of the fledgling stars inside. These illuminate the elements present in the Pillars. In this color-coded image, oxygen shines blue; sulfur, orange; hydrogen and nitrogen, green.

> The Pillars of Creation is the most astonishing image from the Hubble Space Telescipe.

Surrounding the nebula is a cluster of some 8,000 recently born stars that are eroding the Pillars with their intense radiation. Like all young star clusters — such as the Pleiades — they will cling together for several million years, until gravity eventually loses its grip on them.

Then each star will head for an independent career — accompanied, possibly, by a family of planets.

The iconic Pillars of Creation, captured by the Hubble Space Telescope

What's the difference between a star and a planet?

I t's simple: stars shine, planets don't (see Chapter 6). Hang on — we hear you say — planets *do* shine in our night sky. But they have no light of their own; they're illuminated by reflected light from their parent star, the Sun.

A star is different. All those tiny, twinkling points of light you see in the sky are actually vast nuclear reactors. They generate their energy by fusing hydrogen into helium, like a slow-burning H-bomb.

Hydrogen was the first element to be created after the Big Bang (see Chapter 12). But it took Cecilia Payne-Gaposchkin, a young astronomer from a small market town in England, to discover that it was the most common element in the universe.

Her doctoral thesis at Harvard, written in 1925, is still considered to be the best PhD work ever submitted. It was revolutionary — and not everyone agreed at the time. After all, the Earth is made of heavy elements like iron, so it seemed bizarre to suggest the stars were composed of hydrogen.

Astronomers eventually accepted that she was correct, and went on to realize that the hydrogen in the core of a star generates the energy that makes it shine. While planets are too lightweight to ignite a central nuclear reactor, stars are so massive that the pressure and temperature at their cores — running into many millions of degrees — trigger nuclear fusion. And that nuclear reaction is the key to creating and sustaining life in the universe.

A planet bathes in the light of its incandescent star, Wolf 1061.

How do we measure the distances to the stars?

The stars above us look as if they are just stuck on the inside of a great black dome. Yet astronomers are very happy to tell you, for instance, that the famous red star Betelgeuse is almost 100 times farther away than its neighbor in the sky, brilliant Sirius.

How can they know?

It's time for a simple experiment. Hold a finger at arm's length, and view it through one eye (shutting the other) — see where it lies against the wall of your room. Now close that eye and open the other. Your finger seems to jump because your perspective has changed. If you measure the jump, and know how far apart your eyes are, you can work out the distance to your finger.

Now replace your finger with a nearby star, and the wall with a background of distant stars. The shifting viewpoint is now provided by the Earth at opposite sides of its orbit around the Sun. If we measure the amount the star appears to move — and we already know the width of the Earth's orbit — then we can work out the star's distance.

This "parallax technique" is easy to describe, but it's fiendishly hard to implement because the stars are so distant that they seem to move very little. Over the course of six months, Sirius seems to shift about the width of a small coin seen at a distance of 5 km.

That's where the amazing European satellite *Gaia* comes in. *Gaia* has been traveling with the Earth since 2014, and its precision telescope has been constantly pinpointing the positions and parallax shifts of stars. As a result, we now know the distances of over one billion stars throughout our Milky Way Galaxy!

On the cosmic scale, Sirius turns out to be a near neighbor.

What is a light year?

Despite its name, a light year does not measure time; it's the way that astronomers talk about *distances* in the universe. Quite simply, a light year is the distance that light travels in one year. Light is the fastest thing there is. In one second, a beam of light zips across 299,792 km of space — almost the distance to the Moon. In one year, it covers almost 9.5 trillion km — so that's one light year.

The bright stars we see in the night sky are dozens, or even hundreds, of light years away. Brilliant Sirius is 8.6 light years from us, while red Betelgeuse is around 700 light years away —

we are seeing the light it emitted at the time the Black Death ravaged Europe in the 1300s.

The Summer Triangle involves an even more extreme range of distances. Altair lies 17 light years away, and brilliant Vega lies at a distance of 25 light years, but Deneb is thousands of times more luminous and lies a staggering 2,000 light years distant. The light we see today has been traveling to us from the time of the Roman Empire.

And an accurate answer to this question — one light year is precisely 9,460,730,472,580.8 km.

Straddling the Milky Way in the left half of this view, the three stars of the Summer Triangle are not as equal as they seem. Seen clockwise from the far left, they are the brightest stars in this image: Deneb, Vega and Altair.

How far is the nearest star?

Head for Proxima Centauri, the third star in the Alpha Centauri system. The main members of this system (Alpha Centauri A and B) are a close pair of stars similar to the Sun. Together they appear as the third-brightest star in the sky after Sirius and Canopus — a lovely sight in the skies of the Southern Hemisphere.

Proxima (Alpha Centauri C) is a runt by comparison. It's a dim red dwarf star — invisible without a telescope — just one-seventh the size of our Sun. It makes up for this by being a very active flare star with many eruptions, and it boasts a planet about the size of the Earth, although the frequent flares from Proxima rule out life on its surface.

At a distance of 4.25 light years, Proxima is our nearest star. But what does that mean in terms of journey time?

If you were traveling by car at 100 km/h, you're looking at around 46 million years — without refreshment stops. A plane would take five million years. And even Pluto's *New Horizons* probe, the fastest spacecraft in our Solar System, would take 54,400 years to reach Proxima when traveling at over 84,000 km/h.

Will the journey ever be possible? A team that originally included Stephen Hawking are on the job already. They're planning a miniature laser-powered robot probe capable of traveling at 20 percent the speed of light. At this speed, the Breakthrough Starshot project could reach the Alpha Centauri system in just 20 years.

At a distance of 4.25 light years, Proxima is our nearest star.

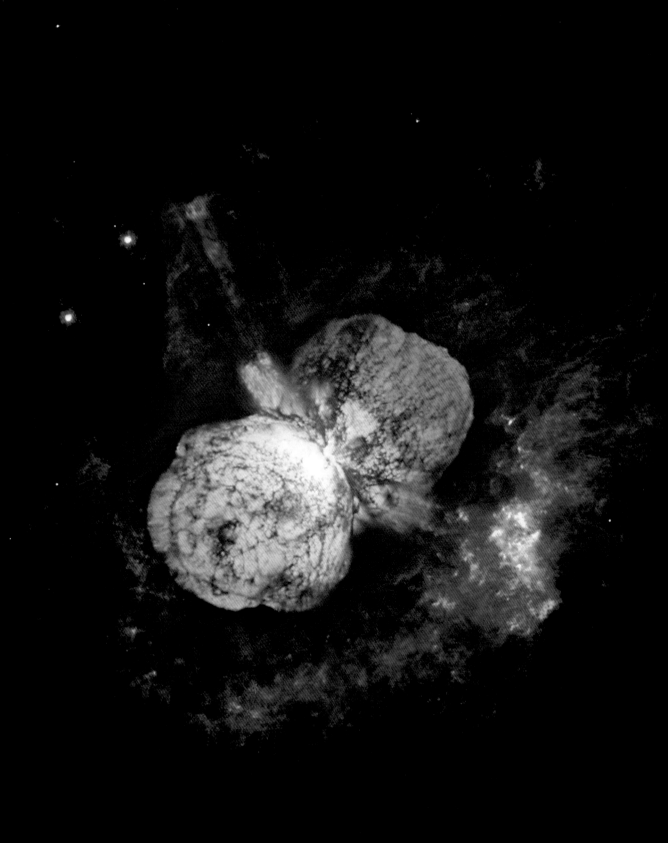

Do the stars we see in the sky still exist?

This is a very common question. People are convinced that many stars don't exist anymore because their light has taken so long to reach us, and that those stars are actually dead.

This idea and the question of stellar distances (see previous question) *really* puzzle adults. The distances and timescales in the universe actually put them off astronomy.

Stars live for billions of years. Our Sun will last for almost 10 billion years.

Are some of the stars dead? Unlikely. Stars live for billions of years. Our Sun, an average star, will last for almost 10 billion years, and much dimmer stars can go on considerably longer. The stars we see are within a few tens, hundreds or thousands of light years away, which means that their light takes less than a few thousand years to reach us. So they're not likely to be dead.

There could be exceptions: stars that have given up the ghost very recently. Very massive, brilliant stars live profligately — they die young as supernovae (see page 196). One suspect is Eta Carinae (pictured here), which is an unstable star famed for behaving erratically. Normally of very average visibility, it flared up in 1843 to become the second-brightest star in the sky.

Since then, it has undergone a number of outbursts, ejecting plumes of gas violently into space. This is all behavior pointing to a future catastrophic detonation, and certain death.

Eta Carinae, visible from the Southern Hemisphere, is a double-star system with a combined luminosity five million times greater than the Sun. The bigger star weighs about 100 to 200 million Suns — an ideal supernova candidate.

Is Eta Carinae still with us? It lies 7,500 light years away, and may be within a few thousand years of its death. Our descendants may soon learn the answer.

The violently unstable Eta Carinae system, captured by the Hubble Space Telescope

How are stars born?

Space isn't empty. Between the stars swims an extremely thin mix of gas and dust — the raw materials of stars-to-be. The dust comes from a previous generation of cool, dying stars. Essentially, it's cosmic soot — microscopic grains of dark material that have wafted off the distended atmospheres of old stars.

Over aeons of cosmic time, the gas and dust clump together to create vast dark clouds. We see them silhouetted against the glowing Milky Way, for example as a dramatic gash in the constellation of Cygnus, and as the dark Coalsack Nebula next to the Southern Cross.

Now gravity can really start to bite. Inside a dark cloud, something is stirring. Knots of dust and gas coalesce, like milk curdling. Under the inexorable force of gravity, the knots shrink, growing hotter as they become smaller and denser. Each of the knots has become a protostar.

Then — a miracle happens. When the temperature at the center of a protostar reaches 10 million degrees, hydrogen atoms collide so fiercely with each other that they create helium.

A nuclear furnace has been switched on, and energy surges out of the protostar, ending its collapse.

In the dark recesses of a cosmic nursery, a star is born.

Out of darkness comes light: a billowing cloud of dense gas and dust in the Carina Nebula

What are the smallest and the biggest stars?

et's go with the little one first, but don't expect a memorable name. EBLM J055-57Ab was recently discovered in a search for planets crossing in front of their companion star. It's around the size of the planet Saturn, but on closer scrutiny it turned out to be a star.

EBLM is a red dwarf star — among the dimmest (and most common) stars known. It weighs in at 85 times the mass of Jupiter, which means that it's just on the dividing line between a star (fueled by nuclear reactions) and a planet.

White dwarfs (see page 194) are even smaller than red dwarfs. No bigger than the Earth, they are merely the collapsed and cooling cores of dead stars, no longer generating any energy.

At the other end of the size spectrum is VY Canis Majoris, in the constellation that's home to Sirius. VY is a red hypergiant, 3,900 light years away from us and 17 times heavier than the Sun. With a cool surface temperature of 3,500°C (6,300°F), this is a star near the end of its life.

Running out of nuclear fuel, VY has become bloated in its declining years. Its size is controversial, but it appears to be over 1,400 times wider than our local star. This means that if placed in the Solar System, VY would extend beyond the orbit of Jupiter.

The red giant seen above, Pi¹ Gruis (π^1 Gruis), is almost a billion kilometers across; to the same scale, the Sun would be smaller than this period. And a white dwarf would be only the size of bacteria.

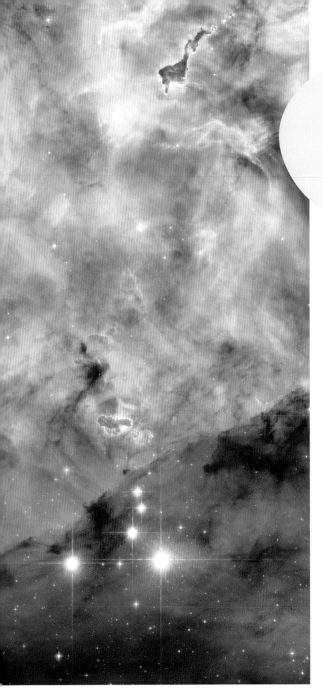

...and what's the brightest star?

The star WR 25 is located within the star cluster Trumpler 16. This cluster is embedded within the Carina Nebula, an immense cauldron of gas and dust that lies approximately 7,500 light years from Earth in the constellation of Carina, the Keel.

Trumpler 16 contains some of the biggest and brightest stars in our Milky Way. But WR 25 is an absolute whopper. It weighs 110 times the mass of our Sun. It's also 6,300,000 times brighter than our local star, making it the most luminous star in our Galaxy.

You can't see WR 25 with the unaided eye; it and its neighboring brilliant stars are embedded in a thick gas and dust cloud. It's circled by another star, though very little is known about this companion.

Lying some 10,500 light years away, WR 25's surface has a staggering temperature of 50,100°C (90,200°F). The star is only two million years old, but it won't have much longer to live. Monster stars like this die in their youth (see page 196).

The brightest star in the Galaxy is WR 25 (white, lower right) in the Trumpler 16 star cluster. The reddish star (lower left) is closer to us, and much less luminous.

Are all stars white?

By no means. Take the constellation of Orion: Betelgeuse glows red, while Rigel shines steely blue-white. Nearby Sirius, the brightest star in the sky, is pure white. Close to Orion, Aldebaran in Taurus shines red. The ultimate red star is Antares in Scorpius; its name actually means "rival of Mars"!

> The ultimate red star is Antares in Scorpius; its name actually means "rival of Mars"!

Arcturus is orange; our Sun and Capella are yellow.

But what do these colors mean? Think of them as taking a thermometer to the stars — you can use them to roughly gauge a star's temperature, though astronomers now have far more accurate ways of taking a star's temperature.

The hottest stars of all — like Rigel, with a temperature of 12,000°C (22,000°F) — glow a fierce blue-white. Lower on the heat scale are the white stars, at around 10,000°C (18,000°F).

Yellow stars come next. Our Sun's surface is roughly 5,500°C (10,000°F), and Capella is slightly cooler, with a temperature of 4,700°C (8,500°F).

Red stars — red giants and red dwarfs — are relatively cool in comparison. They hover around temperatures of 3,500°C (6,300°F). The temperature of one red giant in Canes Venatici (nicknamed "La Superba") can drop as low as 2,500°C (4,500°F).

When you go outside to look at the colors of the brightest stars, don't expect to be dazzled by neon lights. The colors are subtle and best seen through binoculars, but they are there.

A kaleidoscope of star colors lights up the cluster Westerlund 2 in Carina.

How many stars are double?

Surprisingly, about two-thirds of them. Our Sun is rare in being a singleton. Most stars hang together in pairs or even bigger groups — once they've left the nebula where they formed (see page 186).

One of the most famous double stars in the sky is the duo of Mizar and Alcor, in the tail of the Great Bear (or the "handle" of the Dipper or Plough). The pair are quite bright, and are easily visible without a telescope. For a while, controversy raged as to whether the "Horse and Rider" were a genuine pair, or just happened to be lined up in space. Now, with better distance measurements, it appears that they *are* a genuine couple, lying about 80 light years away from us. What's more, they're in a system of six stars, but the others are too faint to be seen with the unaided eye.

Another iconic binary star system is Epsilon Lyrae. It lies next to brilliant Vega in the constellation of Lyra. Look at it with the naked eye, and you'll see that it is double; now look again through a telescope, and you'll find that *both* stars are double! That's why astronomers call Epsilon Lyrae the "Double-Double."

On our doorstep is the Alpha Centauri three-star system. The two main stars are in close binary coupledom — they can approach one another as closely as Saturn orbits the Sun. Circling the first two stars, and at a much farther distance, is red dwarf Proxima Centauri — the closest star to Earth (see page 183).

Undoubtedly, the most beautiful double star in the sky is Albireo, which marks the head of the swan in Cygnus. But *is* it a true binary, or just two stars lined up? The jury is still out.

Whatever the answer, eyeball the pair through a small telescope. You'll discover a beautiful sight: a bright golden-yellow star nuzzling up to a fainter blue, gemlike orb. The Victorian astronomer Admiral William Smyth described them as "topaz yellow and sapphire blue."

The yellow star is aging quickly. It's a massive 70 times heavier than the Sun, and 1,300 times brighter. The blue star is only 230 times more luminous than the Sun, but it makes up for it with its temperature — a searing 13,000°C (23,000°F).

Beautiful Albireo, in the constellation of Cygnus, the Swan.

What's a nova?

Nova means "new," but in celestial terms, it's anything but. Novae are very elderly stars that suffer explosions — but nowhere near as powerful as supernova explosions, that in comparison destroy whole stars (see page 196).

Being the core of a dying star, a white dwarf is *very* hot. The atmosphere ignites and a nova follows.

Novae are binary stars in a tight gravitational embrace. They're obviously old systems, because one of the stars has become a white dwarf (see next question). The dwarf's gravity creates a temporary hydrogen atmosphere when it pulls gas off the outer layers of its companion. Being the core of a dying star, a white dwarf is *very* hot. As the atmosphere builds up, it spontaneously ignites, starting nuclear fusion. Then — BANG! The white dwarf's atmosphere is blasted into space.

It might only be a small amount — perhaps one ten-thousandth the mass of the Sun — but the gas is traveling at thousands of kilometers per second, and it can pack a powerful punch.

The result is a nova — as seen from Earth, the star flares up. The brightening doesn't last long, as all the action is over in a matter of months. And although novae are exciting and offer many insights to astronomers, they're seldom spectacular sky sights.

December 2013 saw the last bright nova to grace our skies — V1369 Centauri. One of the most luminous in recent years was 1975's Nova Cygni, which rivaled Deneb in brightness.

It's estimated that there are 30 to 60 novae in our Galaxy every year, and many have been known to be recurrent; alas, they only repeat the cycle every 100,000 years or so.

One exception is RS Ophiuchi, pictured here. Five thousand light years away in the constellation of Ophiuchus, it goes nova every few decades.

Nova! A white dwarf (right) erupts after seizing matter from its companion. In this artist's rendering, the ejected gas has formed into spiral dust lanes.

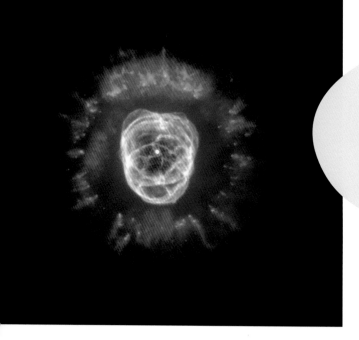

Do stars live forever?

No. Like us, they are born, they live and they die.

Some stars *do* make a bid for immortality. Around three-quarters of all the stars in our Galaxy are red dwarfs. They have temperatures of less than 4,000°C (7,200°F), and weigh just a fraction of the mass of the Sun. Proxima Centauri is our nearest red dwarf, yet it's only visible with a telescope; the majority of red dwarfs are hard to find.

They're powered by nuclear reactions that proceed very slowly, which means that in theory they can live for trillions of years. Some might last for longer than the present age of the universe: 13.8 billion years. Red dwarfs are feisty — and dangerous. Some are covered in star spots, and many boast powerful flares. But in the end, a red dwarf *will* die when its nuclear reactions cease.

What about middleweight stars, like the Sun and Alpha Centauri? Like all stars, they're powered by nuclear reactions, starting with hydrogen being fused into helium in their incredibly hot cores. Stars slightly heavier than the Sun can turn helium into carbon, but that's as far as it goes.

When the nuclear reactions in a middling star pack up, its core collapses and grows hotter. This causes the star's atmosphere to expand, turning it into a red giant.

A red giant star is notoriously unstable. Its atmosphere billows in and out, until in the end, the star puffs it off altogether. The result is a planetary nebula (nothing to do with planets; William Herschel thought they *looked* like Uranus, which he discovered in 1781).

These portents of star-doom are among the most beautiful sights in the sky. Take the Eskimo Nebula: his "parka" is gas streaming away from the dying star and measures two-thirds of a light year across.

Planetary nebulae disperse in a few thousand years, leaving the old star's core exposed to space. The core becomes a white dwarf star — a shrunken corpse devoid of nuclear reactions.

A white dwarf — like the Pup, the faint companion of the Dog Star, Sirius — weighs two-thirds as much as the Sun, but is compressed to only the size of the Earth. And its heat is leaking away into the cosmos. Its destiny is to become a cold, black cinder.

Curtains for the average star: the Eskimo Nebula

What is a pulsar?

I was November 1967, and a young researcher at Cambridge University couldn't believe her eyes. As Jocelyn Bell watched the paper chart emerging from her radio telescope, signals from space were kicking the recording pen across the paper in regular pulses, just over a second apart.

She and her team were baffled. Was it an alien species trying to communicate? Only half in jest, they labeled the chart "LGM-1" — the first signal from Little Green Men.

We now know that such "pulsars" are actually very dense, tiny stars — smaller than London or New York, but containing as much matter as the Sun. They also go by the name "neutron stars," as they're mainly made up of subatomic particles called neutrons.

The pulsar is emitting beams of radio waves, corralled by its powerful magnetic field. As the star spins round, these beams sweep past the Earth and we pick up regular pulses — just as a sailor sees flashes of light from a lighthouse as its lens rotates.

But how can a star end up so small, dense and packed with energy? The clue came from an enigmatic cosmic beast, the Crab Nebula (so named by a Victorian astronomer as it vaguely resembles a crustacean). This glowing gas cloud (seen here) is the mortal remains of a supernova that the ancient Chinese saw exploding in AD 1054. And, right at the heart of the Crab Nebula lies a frenetically spinning pulsar that rotates 30 times every second.

So here's the picture: when a star explodes (see next question), its core collapses down to a tiny ball made entirely of neutrons, whirling around so that it appears to pulse in our skies.

But that's not quite the end of the story. A neutron star's gravity is unimaginably powerful: you'd expend more effort climbing a one-centimeter mound here than you'd use climbing Mount Everest on Earth. And the more mass you cram into a neutron star, the stronger the force becomes. If the original supernova's core was heavier than three suns, gravity becomes an irresistible force.

Instead of settling down as a pulsar, the supernova's core collapses totally to create the weirdest of Nature's monsters — a black hole.

A pulsar forms the beating heart of the Crab Nebula.

Why do stars explode?

A supernova is the biggest bang in the universe since the Big Bang that created it all. It marks the death of a massive unstable star.

A heavyweight star rips through its hydrogen in a relative instant on the cosmic scale — a few million years. When its central fuel supply runs out, just like a medium-weight star (see page 194), it expands to become a red giant — though, as befits its status, the massive star swells into a truly vast monster like Betelgeuse or VY Canis Majoris (see page 188).

But a massive star has the gravitational power to squeeze its core to ever-higher temperatures. Instead of stopping at carbon, the star's nuclear furnace keeps springing back to life, converting the elements in the center to successively heavier products. The core ends up layered like an onion: within a shell of carbon you'll find successive shells of oxygen, neon, magnesium, sulfur and silicon.

When the silicon in the center fuses into iron, the star is in big trouble. In terms of nuclear reactions, iron is the most stable element — you can't fuse iron atoms to create energy. Gravity now wins out. The core collapses completely. The effect is pandemonium. As the core's tempera-ture rises to 5 billion °C (9 billion °F), the iron disintegrates, undoing in seconds what the star has taken millions of years to achieve.

Unsupported, the outer layers start to cave in, only to be hurled out again by a raging torrent of subatomic particles — neutrinos produced in the central inferno. The exploding gases glare

The Spaghetti Nebula (Simeis 147) is the remains of a star that exploded 40,000 years ago.

billions of times brighter than an ordinary star in a supernova explosion that often outshines its own galaxy.

Astronomers have seen more than 10,000 supernovae in distant galaxies. Old records tell of a dozen brilliant supernovae seen in our Galaxy over the past two millennia. And the Milky Way is littered with hundreds of supernova remnants — the expanding fireballs from supernovae that exploded before anyone wrote down what they saw in the sky — like the beautiful Spaghetti Nebula seen here.

BLACK HOLES 10

What is a black hole?

"B lack hole ahead!" It's the scariest warning you could ever hear. If a future spaceship fell into a black hole, it wouldn't just be wrecked; the craft and all the crew would be dragged from our universe and reduced to the atomic level.

And the worst news is that black holes aren't just science fiction; they are science fact. Astronomers have proof positive of black holes in our universe, from monsters a few times heavier than our Sun to cosmic Godzillas billions of times more massive.

The smaller black holes are born when the core of a supernova (see page 195) collapses completely. We find the biggies in the middle of galaxies, where huge amounts of matter have condensed.

A black hole of any size is marked by just one fatal characteristic. It's a region of space where gravity is so strong that nothing can escape — hence it's the ultimate "hole." A black hole can even trap a speeding beam of light, which is why it appears "black."

It's about as bizarre as anything in the universe can get. A black hole "teaches us that space can be crumpled like a piece of paper into an infinitesimal dot, that time can be extinguished like a blown-out flame," according to American physicist John Archibald Wheeler, one of the first to calculate what happens when gravity goes wild.

But there's a bit of a mystery about who coined the phrase "black hole." Wheeler originally just called it a "gravitationally completely collapsed object." The finger seems to point to a fellow astrophysicist, Bob Dicke. His favorite phrase — which he employed when anything was lost around his house, before it came up at scientific meetings — was, "It's been sucked into the Black Hole of Calcutta." That nineteenth-century Indian prison has long been a metaphor for any small, very crowded place you can't escape from.

Computer graphic of a black hole too close for comfort

Who first predicted black holes?

It wasn't Stephen Hawking — nor even Albert Einstein. The honor goes to an eighteenth-century English clergyman, the Reverend John Michell. He was vicar of Thornhill Church in Yorkshire — a beautiful building that houses some of the finest medieval stained glass in the county. Prior to Thornhill, Michell had been Professor of Geology at Cambridge, where he correctly predicted that earthquakes move in waves.

When he got married, Michell left Cambridge for Thornhill. As well as his clerical duties in Yorkshire, Michell also immersed himself in science. He studied magnetometry, optics, gravitation and astronomy — just a few of the subjects at his fingertips.

Out of his astronomical research came Michell's *pièce de résistance*.

In a paper read to the Royal Society in 1783, Michell told his audience: "If there should be in nature any bodies . . . whose diameters are more than 500 times that of the Sun . . . their light could not arrive at us." Michell's "dark stars" were an astonishing prediction of black holes.

But it took over two centuries for the concept to be accepted by astrophysicists. Why? Let the American Physical Society have the last word:

"He was one of the most brilliant and original scientists of the time. Michell remains virtually unknown today, in part because he did little to develop and promote his own path-breaking ideas."

Thornhill Church, Yorkshire, England — birthplace of black holes

How can you detect a black hole if it's completely dark?

With great difficulty! By definition, the culprit does not emit any light or other radiation, so you're not going to find it advertising its presence. Even if you could illuminate it — as a planet is lit up by its sun — the black hole would simply swallow up the light. And black holes born from the death of a star may be mighty, but they are tiny — maybe just a dozen kilometers across.

So, searching for a black hole in the darkness of the universe is like looking for a black cat in a cellar on a dark night. But you have a fair chance of finding your feline if it's drawing attention to itself by having a spat with a white cat. And so it is in the cosmic cat hunt — it's not too difficult to find a black hole that's orbiting an ordinary star we can see, if it's creating a spitting fury by dragging gas from the helpless victim.

A pioneering satellite called Uhuru first picked up signs of a black hole tangling with a star in 1971, when it found X-rays streaming from the constellation Cygnus (the Swan). Here, astronomers discovered a massive star that's being swung round and round by an invisible object that is tearing away the star's matter to form a superhot disc that emits X-radiation. Such a powerful but unseen object must be a black hole. (In case you're wondering why the black hole hasn't gobbled up the entire star, its orbital speed keeps it at a safe distance.)

Cygnus X-1 is as heavy as 11 suns. We've now tracked down around 20 black holes orbiting stars in our Galaxy, and they range from 4 to 26 times the mass of our Sun.

Recently, astronomers have discovered distant black holes spiraling closer together until they merge — recognized by their distinctive "chirp" of gravitational waves (see Chapter 13). And the hearts of distant galaxies contain the most massive black holes of all.

The black hole Cygnus X-1 lies at the center of a disc of brilliantly hot gas that's ripped from its hapless companion.

Will the black hole in the center of the Milky Way swallow us up?

Black holes have gotten bad press as the ultimate in death and destruction — and we admit we're not immune to a bit of colorful language when describing them. But they do have one redeeming feature: you have to be *really* close to a black hole before it will swallow you up wholesale.

So, the short answer to this question is "no." The Milky Way does have a brute of a black hole at its core — known as Sagittarius A* — but we are a safe 26,000 light years out. As the Sun travels around the Galaxy, its orbital speed of 720,000 km/h keeps us from falling toward the central black hole.

That's a good thing, because this black hole is as heavy as four million suns. Astronomers have calculated its gravity — and hence its weight — by watching stars that orbit up close and personal to Sagittarius A*. At its nearest point, the star S2 is the fastest natural object known, whizzing past the black hole at almost 20 million km/h.

The mysterious heart of the Milky Way, as revealed by its output of X-rays. The black hole lurks in the brightest region at center.

What's the biggest black hole?

The most massive black holes are found in the centers of galaxies (see the next chapter for all the lowdown on these star cities). Generally, the bigger the galaxy, the heftier the black hole at its heart.

The black hole that lurks at the core of the Milky Way is an impressive four million times heavier than our Sun. But, on the cosmic scale, it's a mere bantamweight. Take a look at the picture here. The galaxy M87, in the constellation Virgo (the Virgin), is a giant compared to the Milky Way. And it has a supermassive black hole to match — as heavy as seven *billion* Suns.

But this monster is outclassed by the black hole in the galaxy OJ287. Here we find a behemoth weighing in at 18 billion Suns — so powerful that its gravity is holding another supermassive black hole in orbit. The two will merge in about 10,000 years' time.

To top them all, how about a black hole as massive as 66 billion Suns? The current heavyweight champion lies in the center of the galaxy TON 618, named after the Tonantzintla Observatory in Mexico, where it was discovered in 1957. This black hole is heavier than an entire medium-sized galaxy, like the Milky Way's neighbor the Triangulum Galaxy!

Galaxy M87 — its distinctive jet of blue glowing gas is ejected by activity near its central supermassive black hole.

What would happen if I fell into a black hole?

It would be very uncomfortable . . . like being ripped apart.

Heather was on the UK's premier TV news program with an interviewer who was being condescending. "What would happen if I were to fall into a black hole?" he asked.

She grabbed her water glass. "This is a black hole," she told him. "Now imagine heading toward it, feet-first. The gravitational pull on your feet would be so much greater than that on your head that you'd be stretched into a long, thin tube. It's a process known to astronomers as 'spaghettification.' Oh — and you'd be torn up."

This would be the fate of an unwise astronaut facing a "stellar mass" black hole. These light-

weight black holes are the result of a supernova explosion. Their gravity creates deep pits in the fabric of space, from which there is no escape.

There's another way forward for our intrepid black hole explorer: pick a biggie. The black holes commanding the centers of supermassive galaxies weigh in at millions, or even billions, of times the mass of the Sun. These gentle giants make much shallower gravitational "wells" in the warp and weft of the cosmos.

An infalling astronaut would experience only moderate spaghettification forces as he or she fell in. And now the real adventure is about to begin . . .

> These lightweight black holes are the result of a supernova explosion. Their gravity creates deep pits in the fabric of space.

Anyone for spaghetti?

What happens inside a black hole?

Okay, suppose you survive the spaghettification. Now you are dropping near to the edge of the black hole — the event horizon. (It's called that simply because the black hole's gravity prevents events happening inside from being seen by anyone outside.)

Now let's introduce a bit of scientific weird. Our modern theory of gravity is called the General Theory of Relativity (proposed by Albert Einstein in 1916), and, when it comes to black holes, the emphasis is *relativity* — everything is relative to where and when you're observing it.

Imagine Uncle Albert is on a spaceship at a safe distance, watching you freefall. He sees gravity distorting time as you fall in. With a powerful telescope, he sees the seconds on your watch tick slower and slower. If you worked out you would fall through the event horizon at 12 noon, Albert would see your watch slow down so much it would never quite reach midday. As your time seems to dawdle, he would observe you falling more and more slowly, so that you never quite reach the event horizon.

For you, it's much simpler: you'd fall through the event horizon at 12 noon, without really noticing it. (There's a recent theory that you'd hit an incandescent "firewall" that would frazzle you, but scientists are still arguing about that one.)

Once inside the black hole, you'd see a distorted view of brilliant stars and galaxies above you as their light falls toward you and is amplified by the strong gravity. The black hole is empty — except for a tiny speck at the center where all the stuff that's previously fallen in has accumulated. Theoretically speaking, this "singularity" has zero size, and it's infinitely dense.

Before you know it, you hit the singularity and all of you is squashed down to vanishing point.

Computer graphic of the frenetic activity around a black hole: whatever happens inside the dark sphere of the event horizon can never be seen from the outside.

Can you travel through a wormhole?

Hang on, you may be saying, we've been painting black holes as the ultimate — and lethal — dead end; but there are plenty of sci-fi movies in which spaceships simply whiz into black holes, then out again somewhere else, maybe in another universe.

The idea that a black hole can be a shortcut through space actually goes back to Einstein in the 1930s. But three decades later, John Archibald Wheeler proved you couldn't get through without being crushed at the singularity.

Back to the drawing board . . . what about a black hole that's spinning around? Now, the singularity isn't just a point, but a ring, and at first, the ring looks like the gateway to another cosmos. In the 1980s, though, black-hole doyen Kip Thorne proved it was highly unstable — try to pass through, and it would collapse on you.

That was unfortunate, because the great astronomy communicator Carl Sagan had asked Thorne how the heroine in his book *Contact* could travel instantly to a planet across the Galaxy. Black holes clearly wouldn't do. On a long drive down Interstate 5 in California, Thorne let his mind freewheel. He realized that a black-hole bridge could be kept open if you shored it up with some kind of "exotic material" that has an antigravity effect — pushing where gravity pulls.

This is a called a wormhole — a shortcut through space that lets you travel either way, and arrive instantly at another planet or even a distant galaxy. Wormholes are still speculative, but they've inspired some great science fiction: *Contact* — book and movie — and then films like *Stargate*, *Timeline* and *Interstellar*, not to mention space-hopping episodes of *Star Trek*.

Though it's purely theoretical, the mouth of a wormhole may provide a safe passage to distant parts of the cosmos — or even to another universe.

THE MILKY WAY AND OTHER GALAXIES

11

To the Romans, it was *Via Lactea* — the Path of Milk. And on a clear, transparent night, the glistening arch of the Milky Way lives up to its name in every respect.

The ancients had a delightful legend as to how the Milky Way came to be. Juno, the long-suffering wife of philandering Jupiter, was tricked into nursing yet another of his illegitimate children, born to the mortal Alcmene. Baby Hercules — famed for his 12 labors — started off in the way he was destined to continue.

The boisterous child tickled Juno's nipple; the milk spurted past Hercules' mouth and into the sky. Being born of a mortal woman, and missing out on a goddess' milk, Hercules' future was sealed. He would be forever mortal. But at least we have the Milky Way to thank him for.

Apologies for coming back to reality, but the Milky Way isn't a stream of celestial milk — as Galileo found out when he swept it with his humble telescope. He described it as "but a congeries of innumerable stars."

Even when viewed through modest binoculars, the Milky Way is a glorious sight. Stars, glowing nebulae, pitch-black clouds, star clusters that look like collections of precious stones — this panorama is the Milky Way (see images on pages 3, 27 and 182).

Welcome to our Galaxy. We live in a city of some 200,000 million stars, of which our Sun is but one. It's an average, middle-aged star living in the suburbs of a vast stellar conurbation. The stars we see scattered across the sky are our near neighbors; most live no more than tens or hundreds of light years from us. But our Galaxy is lens-shaped. In the immortal words of the late Sir Patrick Moore, it looks like "two fried eggs, clapped together back-to-back."

Perspective makes the more distant stars in our Galaxy appear to crowd into a band — the arc of the Milky Way. Sweep it again with binoculars or a small telescope, concentrating on the constellations of Scorpius and Sagittarius, and you'd be looking at our Galaxy's "yolk," where all of its most beautiful sky sights are congregated.

In mythology, the Milky Way gushed from the breast of the goddess Juno, as depicted here by Jacopo Tintoretto around 1580.

Does the Milky Way have a twin?

I t could be a postcard photo of the Milky Way. NGC 6744 lies 30 million light years away, and astronomers agree it's a dead ringer for our own Galaxy.

Living, as we do, in the disc of a spiral galaxy, it's hard to work out the structure of our own star city. But astronomers have used multi-wavelength techniques to map the Milky Way, and have a pretty good idea of how it looks.

It takes our Sun roughly 225 million years to circle the Galaxy — a period called the "galactic year."

At 175,000 light years across, NGC 6744 is bigger than the Milky Way (our Galaxy is 100,000 light years in diameter). And, like the Milky Way, it has a bright companion galaxy (lower right) that looks uncannily like the Large Magellanic Cloud (see page 218).

If this were the Milky Way, the Sun and Solar System would lie halfway from the center, in a "spur" of stars between two major spiral arms. These stunning appendages are the scene of the major action in a spiral galaxy — the arms are packed with wraiths of dust and gas from which new stars are born. The bright knots in the magnificent arms are stellar nurseries.

The center of NGC 6744 is made of old stars — the first in the galaxy to be born. Take a look at it closely and you'll see that it's elongated. Like the nucleus of the Milky Way, it's made of a bar of stars, which is why both galaxies are classified as "barred spirals." Most spiral galaxies have a bar — probably a result of gas being channeled into their central regions and triggering starbirth long ago.

Spiral galaxies, like the Milky Way and NGC 6744, spin around, just as the planets revolve around the Sun. It takes our Sun roughly 225 million years to circle the Galaxy — a period called the "galactic year." So far, our Sun has clocked up 20 galactic years. And as for us human beings — we've managed just one thousandth of a circuit of our Milky Way.

Peacock galaxy: NGC 6744 in the constellation of Pavo

What's the closest big galaxy?

It has to be said that the autumn skies of the Northern Hemisphere are *boring*. Gone are the striking constellations of summer, and it will be a while yet before the dazzling winter constellations take center stage.

The heavens are awash with "wet" star patterns — straggly faint groupings like Aquarius, Pisces, Hydra and Cetus. Just above these denizens of the deep is the uninteresting constellation of Pegasus. It's a large empty square of stars, and protruding out of one corner is a line of three rather anemic-looking stars. This is the constellation of Andromeda.

But look more closely. Just above the middle star is a fuzzy patch; even in light-polluted areas, you should be able to spot it. This is one of the glories of the northern sky: the Andromeda Galaxy.

Lying 2.5 million light years away, this beautiful galaxy lords over our small galaxy cluster, the Local Group, which is home to some 50 galaxies.

Andromeda and the Milky Way are by far the largest.

The Andromeda Galaxy is huge — 220,000 light years across (twice the size of our Galaxy). If you look at it through a small telescope, you'll discover that it covers an area four times the size of the Full Moon.

This colossus of a galaxy is surrounded by a family of at least 14 dwarf companions, one of which is visible in our image (bottom).

Like the Milky Way, Andromeda is a spiral galaxy. But it's angled so steeply to us that we miss out on seeing many of its glories. We do know that it contains a trillion stars (as compared to our meager 200 billion). And its core is home to a supermassive black hole, which weighs as much as 100 million Suns.

The Andromeda Galaxy has a little surprise in store for us in the future. You'll find it revealed at the end of the chapter.

The glorious Andromeda Galaxy: gem of the northern skies

Are all galaxies spiral-shaped?

Not at all — it's just that spirals are very photogenic. Our picture features a lenticular galaxy. The Sombrero Galaxy and its siblings are closely related to spirals, but they have no prominent spiral arms.

Not surprisingly, this galaxy has loads of fans! You can spot it through an amateur telescope, and even through good binoculars. Fifty thousand light years across, the Sombrero lies some 30 million light years away in the constellation of Virgo. It's renowned for its sensational dark rim of cold dust and gas, where stars are being born. This rim and the huge bulge of stars within give the Sombrero its name. The galaxy's nucleus harbors a supermassive black hole weighing in at a billion Suns. Surrounding the central bulge is an immense halo of 1,000 to 2,000 globular clusters of old red stars.

Elliptical galaxies — the next group — also have no spiral arms. The smallest, dwarf ellipticals, are one-hundredth the mass of the Milky Way. Giant ellipticals, on the other hand, can boast ten *trillion* stars or more. The biggies got that way by colliding and merging with each other. The largest known galaxy, IC 1101 (in the constellation of Serpens), contains 100 trillion stars and is a whopping six million light years across. That's 60 times wider than the Milky Way.

What all ellipticals have in common is that they're largely red and featureless; they lack the gas that fires starbirth.

Not so the irregular galaxies, which are brimming with stellar raw materials (see page 218). These shapeless little runts of the cosmos typically weigh one-tenth of the Milky Way, but it's estimated that they make up a quarter of all the galaxies in the universe.

Irregulars, with their star-forming regions, can be very pretty, but there's not much point in being pretty if you're going to be consumed by a giant galaxy — as is so often the case.

The aptly named Sombrero Galaxy

What makes galaxies violent?

Violence takes place in galaxies for two reasons. The lesser form of galactic violence is a sudden surge of starbirth — that's what you find in a "starburst galaxy." Starbursts can be triggered by a gravitational interaction with a neighbor, or the occasional episode of galaxy-gobbling.

The classic starburst galaxy is the sensational, ragged-looking M82, visible through binoculars or a small telescope in Ursa Major. Its bigger sibling, the serene spiral M81, wreaks gravitational havoc on its neighbor. Whenever their orbits bring the two galaxies close together, gas from M81 is funneled into the core of M82, triggering a starburst. The rate of starbirth rises to 10 times that of the Milky Way, sometimes more.

Far more violent are galaxies that have enormous active black holes at their centers. One such "active galaxy" is Centaurus A. As you can see in this image, Centaurus A is a mess. It's a supergiant elliptical galaxy feasting on a spiral galaxy, and gas and dust have been so compressed that stars are forming here. But what makes the galaxy far more violent is a black hole at its core weighing in at 55 million Suns.

This black hole is feeding on gas from the cosmic banquet. It chucks out all kinds of radiation, including radio waves. In fact, Centaurus A was one of the first "radio galaxies" to be discovered.

Astronomers using radio telescopes (see Chapter 4) have now found thousands of powerful radio galaxies. These violent denizens of the cosmos are giant ellipticals, with central supermassive black holes that shoot jets of electrons into space. When the jets hit the intergalactic medium, they create enormous clouds straddling the galaxy.

One of the most powerful radio galaxies is Cygnus A. It boasts an active black hole weighing 2.5 billion Suns, and jets that travel close to the speed of light.

One of our very favorite radio galaxies — Centaurus A

What are the Magellanic Clouds?

The two Magellanic Clouds are a glorious sight in the Southern Hemisphere. They look like detached portions of the Milky Way, but they're independent galaxies in their own right.

From Africa to Australia, our very distant ancestors knew them well. To the seafaring Polynesians, they were important navigational beacons. The name we use today was probably bestowed by the scholar Antonio Figafetta, who accompanied Ferdinand Magellan on his circumnavigation of the globe in 1519.

Until recently, the Magellanic Clouds were thought to be the Milky Way's nearest neighbors. But with the discovery of two tiny galaxies closer in, they've lost that status. Nevertheless, they *are* close on the celestial scale. The Small Magellanic Cloud (SMC) lies 200,000 light years away; the Large Magellanic Cloud (LMC) is even closer, at 160,000 light years.

Both galaxies used to be classified as irregulars. But new observations have made astronomers reclassify the LMC as a "Magellanic spiral." It has a bar of stars, proving that it was once a spiral galaxy; like the SMC, it has been ripped apart. And gravity has pulled out a huge ribbon of gas — the Magellanic Stream — between the two galaxies and the Milky Way.

Both Clouds are small compared to our Galaxy. The LMC is 14,000 light years across, and contains 10 billion stars; the SMC is half its size, and home to seven billion stars. Each is packed with raw materials destined to form

The Magellanic Clouds soar over the ALMA radio telescope array in Chile.

billions more stars. Among their many nebulae, the chief glory is the Tarantula Nebula in the LMC. It's 600 light years across. If we were to replace the famed Orion Nebula in our Galaxy with the Tarantula, it would appear 25 times wider — and bright enough to cast shadows.

For all their proximity, there's much we *don't* know about the Magellanic Clouds. Are they satellites orbiting the Milky Way? Probably not — astronomers think they're traveling too fast. Will they ever smash into our Galaxy? Possibly . . .

What are galaxy clusters?

Galaxies are gregarious, and it's all down to gravity. Our Milky Way is no exception; it has teamed up with other galaxies to create the Local Group. The Andromeda Galaxy is the biggest; next comes the Milky Way and then the Triangulum Galaxy. The other galaxies are mostly insignificant dwarfs.

The Local Group is typical of groups of galaxies. Not more than 3.5 million light years across, it struggles to reach a membership of 50.

Not so galaxy clusters. They contain hundreds to thousands of galaxies, and range in size from 6.5 million to 62.5 million light years across. These are cosmic giants, reaching masses between 100 and 1,000 trillion Suns.

It's amazing that most of this mass is not in the form of stars. They account for just 1 percent, while 9 percent consists of gas between the galaxies. The remaining 90 percent is invisible dark matter (see Chapter 12 for the lowdown on this mysterious cosmic entity).

The biggest galaxy clusters are populated almost entirely by red giants, living out their twilight years in elliptical galaxies. Most often, the centers of clusters are dominated by giant ellipticals, which grow fat by gobbling smaller galaxies.

Galaxy clusters are the largest gravitationally bound agglomerations in the universe. They were thought to be its biggest structures until the 1980s, when superclusters — clusters of clusters — claimed the title. Now we know that superclusters link into huge filaments of galaxies that can stretch over hundreds of millions of light years; the filaments enclose empty voids, so the universe on the largest scale looks a bit like Swiss cheese.

Nearby galaxy clusters include Virgo, Fornax, Hercules, Coma and Norma. The Norma Cluster lies near the center of the Great Attractor — a supercluster whose mass can affect the expansion of the universe.

If you want to find a nearby cluster, look no further than Virgo, home to the Virgo Cluster of 2,000 galaxies, some 54 million light years away. More than that, it's the center of our Local Supercluster, which includes the Local Group and many other small groups of galaxies. Sweep the Y-shaped bowl of Virgo with a small telescope. You'll pick out many faint galaxies — star cities with which we share our destiny.

One of the biggest galaxy clusters, Abell 1689 lies an immense 2.2 billion light years away.

What is
a quasar?

The young Dutch astronomer Maarten Schmidt was baffled. It was the early 1960s, and radio astronomy was in its infancy. Scientists were beginning to understand what kinds of objects emitted powerful radio waves. But there were some loose ends, such as a star in Virgo, visible in amateur telescopes. 3C 273 was chucking out radio waves, but stars generally don't produce radio emissions. So Schmidt — with the iconic Hale Telescope on Palomar Mountain at his disposal — checked its distance.

A quasar is the brilliant disc of gas, circling a supermassive black hole at the heart of a galaxy.

By spreading its light into a spectrum (see next question), Schmidt got a result of two billion light years — far more distant than any galaxy known at the time. Other "radio stars" swiftly followed. Puzzled astronomers called them quasi-stellar radio sources, which mercifully got shortened to quasars.

Now we've discovered some 200,000 quasars, and we know what they are. A quasar is the brilliant disc of gas — an accretion disc — circling a supermassive black hole at the heart of a galaxy. The heaviest of these black holes can weigh the same at dozens of billions of Suns.

Quasars are incredibly luminous. The accretion disc in a typical quasar boasts the brightness of a thousand Milky Ways — all coming from a region the size of our Solar System. Because the

DANGER! Quasar ahead . . . This artist rendering depicts the distant quasar, ULAS J1120+0641.

quasar is so bright, its light tends to drown out the host galaxy (though powerful telescopes like the Hubble can pick these out).

And while black holes suck in matter, the frenetically spinning accretion disc spits matter and energy out. The most powerful quasar is SDSS J1106 + 1939. Its power output is some two billon times that of the Sun.

You've probably noticed that our description of a quasar is similar to our description of radio galaxies (see page 217). In fact, astronomers think they are very similar, except that in the radio galaxy the bright accretion disc is hidden.

Quasars are denizens of the distant past; by now, their energy has been spent. Because their light has taken billions of years to reach us, we see quasars as they were when they were very young. They're the nearest we can get to time travel, shining a spotlight on the early universe.

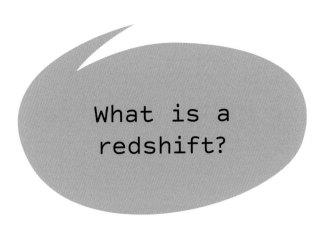

What is a redshift?

In the 1910s, American astronomer Vesto Slipher began a forensic investigation of the faint spiral-shaped objects in the sky (like Andromeda) using a spectroscope that spread out their light into a spectrum of colors. His boss, Percival Lowell (of Canals on Mars fame, see Chapter 14), was convinced these "spiral nebulae" were nearby solar systems in formation.

The spectra were crossed by lines created by various elements (see Chapter 13), and Slipher knew that the position of the lines would shift if the object was moving. This is the Doppler Effect — it's the reason why an ambulance siren has a higher pitch as it comes toward us, then drops abruptly as it speeds away. It's the same with light. If an astronomical object is coming toward us, all its wavelengths are shortened and shifted toward the blue end of the spectrum. If it's receding, the wavelengths are lengthened and move toward the red end of the spectrum; they suffer a *redshift*.

By 1923, Slipher was amazed. Of the 43 spiral objects he'd examined, almost all showed a redshift; they were rushing away from us — the fastest at an unprecedented 6.5 million km/h. By this time, he also believed that the "spiral nebulae" were actually distant star cities, cousins to our Milky Way.

In 1927, a Belgian priest named Georges Lemaître realized that the most distant of these galaxies were rushing away fastest, and that the whole universe was expanding. Unfortunately, his paper "A homogeneous universe of constant mass and growing radius accounting for the radial velocity of extragalactic nebulae" was published in a Belgian journal that most astronomers didn't read.

That's where the self-promoting American astronomer Edwin Hubble comes in. True, he did measure the first accurate distances to other galaxies — and in fact was the first to use the word "galaxy." But in the 1920s, he stole Slipher's redshifts without any credit, and two years after Lemaître, Hubble loudly proclaimed the link between distance and redshift. As a result, it's now known as "Hubble's Law" — and of course he now has a space telescope named after him.

Whatever the history, the redshift is an immensely powerful tool for astronomers. If you can measure the redshift of a galaxy or a quasar, then you can apply Hubble's Law to work out its distance — right out to objects over 12 billion light years away.

Odd one out: the galaxy at top left of Stephan's Quintet is only one-seventh as far as the others, and has one-seventh the redshift.

Is our Milky Way going to collide with another galaxy?

Galaxies like to dance. The Antennae make a beautiful sight; they are named for the curving arcs of dust and gas thrown out by their merger. In the case of Centaurus A (see page 217), the "dance" is more like cannibalism — the galaxy is devouring a smaller companion.

Most galaxies will undergo a close encounter or a merger in their lifetimes — particularly if they live in rich galaxy clusters. The Milky Way is no exception. It's already interacting with the Magellanic Clouds, creating the Magellanic Stream. But will it be able to cope with the full-on treatment that's coming up?

Our Milky Way is a giant spiral made up of 200 billion stars, including the Sun. Most of the other galaxies are speeding away from us, but there's one exception. The giant Andromeda Galaxy is heading toward us. Astronomers believe that we're headed for the mother of all cosmic traffic accidents — when Andromeda and the Milky Way collide.

It's not quite as dire as it sounds. The stars in the galaxies are so thinly spread that they are unlikely to actually strike one another. It would be like two disciplined armies passing each other in orderly ranks.

But there will be exceptions. Many stars will be whirled around by wild gravitational forces, spinning many of them into long streamers — like the Antennae. Other stars could be flung out altogether.

The gas clouds in Andromeda and the Milky Way will smash into one another, creating a wild frenzy of starbirth (like the pink regions in the Antennae image). Eventually, it will all calm down; our two galaxies will merge together into a huge elliptical galaxy that astronomers are calling Milkomeda.

But don't reach for your accident insurance claim yet — the crash between the Milky Way and Andromeda won't take place for another four billion years.

Galaxies in embrace: the Antennae, in Corvus, are busy making stars.

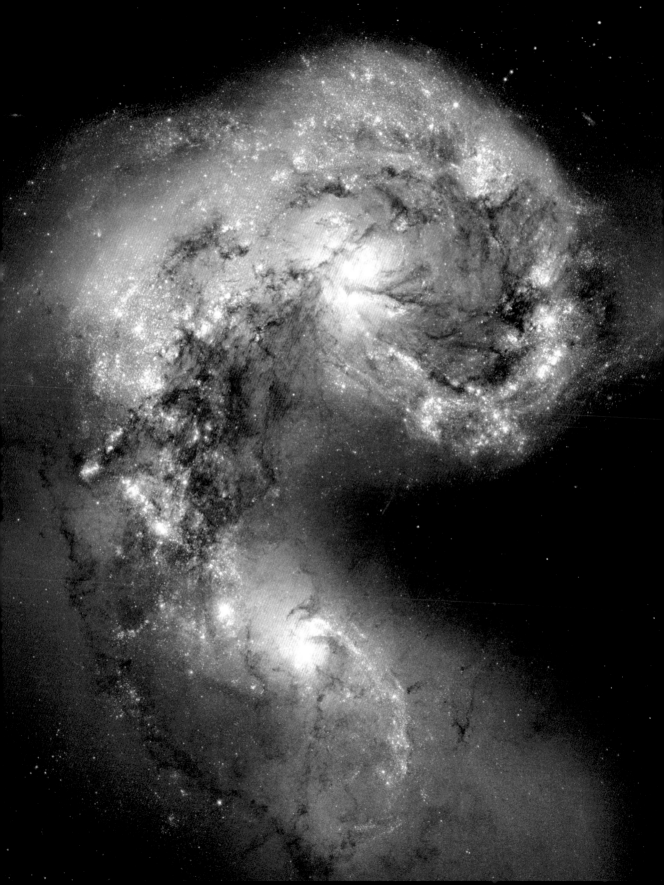

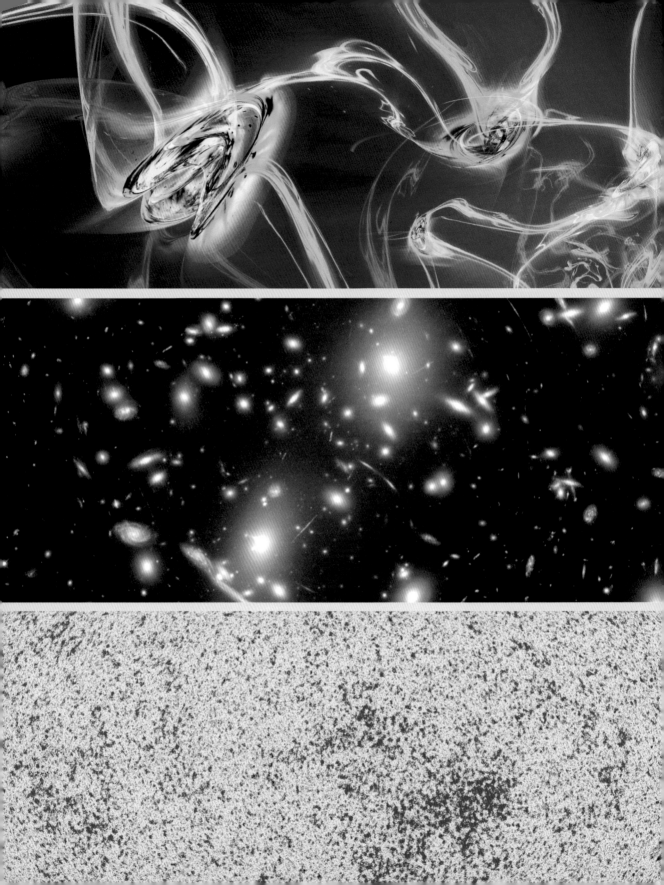

THE BIG BANG 12

What was the Big Bang?

Well, let's start this chapter with a couple of the real BIGGIES. It's impossible for us to chat with anyone for very long before they ask about how the universe began — and why.

Back to basics first. The universe is filled with galaxies — giant star cities, like our Milky Way — that live together in swarms called galaxy clusters. A century or so ago, astronomers discovered that these clusters are racing apart from each other.

The Belgian priest Georges Lemaître realized that the whole universe must be expanding from a tiny point. He named it the Primeval Atom, though now we call it the Big Bang.

There's oodles of evidence that the universe actually began with just a small lump of energy, which then erupted to an enormous size in the tiniest fraction of a second — a period that astronomers call "inflation" — creating vast amounts of energy and matter in the fireball. Since then, the cosmos has been growing inexorably larger.

One theory of how the expanding universe began

...and what caused it?

That brings us back to the ultimate origin. How was everything created, and what happened *before* the Big Bang?

The simplest idea is that there wasn't a "before" the Big Bang; time started in the moment of creation, along with space, energy and matter. Lemaître presciently said: "We can conceive of space beginning with a Primeval Atom, and the beginning of space being marked by the beginning of time." In that case, it does not make sense to ask what caused the Big Bang — it simply happened.

A more modern version says that there was originally "quantum foam" in some other dimension, where energy bubbled up and disappeared again. One of those tiny energy bubbles happened to go into a state of inflation; instead of fading away, it blew up to become our universe.

Here's a much simpler idea. Suppose that the universe is constantly expanding and contracting. In the distant past, the cosmos collapsed down into a Big Crunch, which then bounced back in the Big Bang that made our current universe. The nice thing about this cyclical universe is that it could have been expanding and contracting forever. So you can push the beginning of the whole thing back to the infinitely distant past, and not worry about a beginning at all.

Or perhaps universes procreate . . . In our universe, there are plenty of black holes littering space (see Chapter 10), and in the middle of each of them, a whole load of matter scrunches down to zero size. Where does the dense concentration of energy in this singularity go? According to one theory, it bursts out into another dimension, to spawn a new universe.

People who hold to this theory say that our universe simply erupted from a star or a quasar that previously collapsed in another cosmos. Given that every cosmos creates billions of black holes, there would be a vast family tree of different generations of universe, which could go way back infinitely far in the past.

An artist's impression of the cosmic foam that may have given birth to our universe and others

How can we measure the age of the universe?

Method 1: reverse the movie. We know how fast the universe is expanding, so we can trace the motions of all the galaxies back to the point when they were all on top of each other. Astronomers first tried this in the 1930s; unfortunately, they got the distances to the galaxies so wrong that the universe appeared to be younger than the Earth. Oops! Now, instruments like the Hubble Space Telescope are giving us much better measurements that show the universe is nearly 14 billion years old.

Method 2 is a lot more complicated. Astronomers can pick up the afterglow of the hot gas spewed out from the Big Bang (see page 240), and they find it's mottled like the embers in a barbecue. The size and brightness of these clumps depend on several important things, including the age of the universe and what it is made of (see next two questions). The latest results from the European Planck telescope pin the age down to 13.8 billion years, and that's the most accurate answer we have today.

Method 3: we can do a bit of local archaeology, looking for the oldest stars around us in the Milky Way. Most of them turn out to be about 13 billion years old, which fits well with our Galaxy and its stars forming soon after the Big Bang.

The star we feature in the picture, though, is a rebel. HD 140283 is the oldest star we know, and it's been nicknamed the "Methuselah Star," after the man in the Bible who reputedly lived to the age of 969. This star seems to be 14.5 billion years old — meaning it was born before the Big Bang. Oops again . . . but there is a bit of leeway in the measurements, and the celestial Methuselah could be as young as 13.7 billion years, born immediately after the Big Bang. If we lived on a planet orbiting this star, we would have witnessed the entire history of the universe.

The Methuselah Star has been around to see it all.

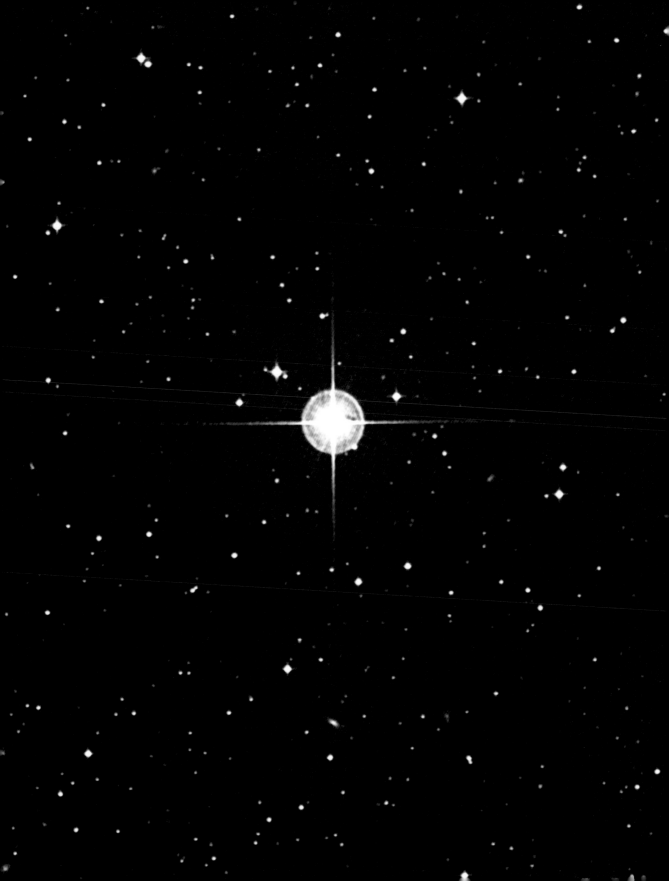

What is dark matter?

"Most of Our Universe Is Missing"
— *Horizon*, BBC

I t's a stunning headline — and it's true. The stars, planets and galaxies we see around us make up only 5 percent of the total "stuff" in the universe. Yes, 95 percent of the universe isn't made of atoms, but consists of unknown, invisible material.

It comes in two types that are totally different but have confusingly similar names. Let's start with dark matter and leave dark energy to the next question.

The story begins in the 1930s with an irascible Swiss astronomer working in California. Fritz Zwicky disliked his conventional American colleagues so heartily that he referred to them as "spherical bastards" — they were bastards whatever direction you viewed them from. And his science was often better than theirs.

Zwicky was investigating galaxy clusters, and he discovered that the galaxies inside were moving so fast they should have escaped by now. He deduced that there was a huge pool of invisible matter within the cluster, and its gravity was penning the galaxies in. He called it "dark matter" — but his reactionary colleagues were having none of this strange notion.

In the 1970s, American astronomer Vera Rubin found that the outer arms of spiral galaxies like our Milky Way are spinning around so fast that their stars should have whirled off into deep space. She deduced that individual galaxies must be packed full of dark matter as well.

Finally, dark matter was brought into the mainstream when astronomers could see its effect directly. In the wonderful image here, we see how the gravity of the dark matter in a nearby galaxy cluster is bending the light of distant galaxies passing through, and distorting them into weird banana shapes.

The conclusion: there's five times as much dark matter in the universe as ordinary matter. We don't know what it's made of, but the smart money is on a kind of subatomic particle that's extremely elusive and only gives itself away by the pull of its gravity.

Gravitational mirages around the galaxy cluster Abell 370

What is dark energy?

ALL MATTER AND ENERGY

All Other
Visible Atoms
0.03%

Hydrogen
and Helium
1.37%

Invisible Atoms
3.4%

Cold Dark Matter
25.8%

Dark Energy
69.4%

Just as astronomers were coming to grips with dark matter (see previous question), nature threw them another curveball.

In the late 1990s, two teams of astronomers discovered that the expanding universe isn't just coasting along — it's growing faster and faster. There must be a force spread over the cosmos that's pushing the clusters of galaxies apart. The researchers could have given this force a comprehensible name like "cosmic repulsion," but instead they chose the mightily confusing "dark energy."

Surprisingly, there's far more dark energy than ordinary matter and dark matter put together.

Oddly enough, 80 years earlier Albert Einstein had proposed a force very like this, to balance up the universe. He named it "the cosmological constant." Later, when astronomers found the cosmos is actually expanding, Einstein took it out of his equations again, calling it his "biggest blunder."

Nobody has the foggiest idea what dark energy is, except that it's a kind of antigravity force. But we know it exists, not just because it's making the universe accelerate, but because it has left its imprint in the radiation from the Big Bang. As we saw above (see page 232), researchers can discover a lot about the universe from the patterns in these ancient embers — not only the age of the universe, but also the amount of dark matter and dark energy. The precise recipe for the cosmos turns out to be 69.4 percent dark energy, 25.8 percent dark matter, and just 4.8 percent ordinary matter.

Surprisingly, there's far more dark energy than ordinary matter and dark matter put together. That's summed up in the pyramid seen here. It shows how little of the universe is visible to us, and how much is hidden in various forms.

Visible matter is just the pinnacle of a great dark pyramid.

Are there more stars in the universe than grains of sand on Earth's beaches?

I n a memorable moment in his TV series *Cosmos,* Carl Sagan picked up a handful of sand from beside the sea and exclaimed: "The total number of stars in the universe is larger than all the grains of sand on all the beaches of the planet Earth."

Is that true? And how can we know?

For a start, a grain of sand is about a quarter of a millimeter across, so Sagan's handful contained roughly 10,000 sand grains. If you fill a seaside bucket, you'll have 100 million grains. That's one-third the population of the United States, but it's a mere nothing on the cosmic scale. The Milky Way contains 200 billion stars. With that number of sand grains, you could build a sand castle as tall as the average adult and as wide as your outstretched arms.

Okay, but even a giant sand castle is small compared to a beach — and think of all the beaches in the world. If you went on a massive building spree, surely you could pile up more sand castles than there are galaxies in the universe, right?

According to NASA, the total coastline of the world is 620,000 km, and around 10 percent is blessed with sandy beaches. That's about 60,000 km of beaches, ranging from tiny coves in the Greek islands to the 250-km-long Praia do Cassino in Brazil.

Now we have to do some educated guesswork. Let's say a beach is about 200 meters wide, and the sand is 10 meters deep. That means the total volume of sand is about 100 billion cubic meters. If you had the time and the energy, you could turn that into 30 billion giant "Milky Way" sand castles.

It's a big number — but less than the 200 billion galaxies that throng the observable universe. So Carl was right: the stars in all these galaxies outnumber the Earth's sand grains.

Is there an edge to the universe?

The answer is yes and no. And we are not just hedging our bets here. Both answers are true, but in different ways.

Imagine we have the ultimate in powerful telescopes: it can peer deeper into space than the Hubble and its successor, the James Webb Space Telescope; in fact, it can see as far as we want to gaze. Can we see forever?

Perhaps surprisingly, the answer is no. The problem is that light takes time to travel through space, so we see distant parts of the universe as they were in the distant past. Once our telescopes are peering out 13.8 billion light years, we are seeing back to the time of the Big Bang. There was nothing before the Big Bang, so we can't see anything farther away. The whole universe we can ever hope to view — the *observable* universe — is contained in a sphere around us, with a radius of 13.8 billion light years.

But we can be a bit more imaginative. Suppose we could somehow stand outside the observable universe, and have a way of seeing everything that exists at the present time. Now the picture is different. Galaxies stretch out in all directions, far beyond the radius of the observable universe.

With our new perspective, surely we'll see an edge to the universe somewhere? But now we have a different problem. If we find an edge, what lies beyond it? Even if it's just empty space, that would be part of the universe too.

So we have to conclude that the universe actually does not have a boundary. It extends out to infinity in every direction.

A medieval explorer checks out the edge of the universe in this 19th century Camille Flammarion woodcut.

What happened to the Steady State theory?

Every so often, along comes a scientist who is a maverick. Not content to work with others in a large, conformist team, he or she resolutely plows a lone furrow. Such a scientist was Fred Hoyle. A blunt Yorkshireman, Fred was often as wrong as he was right. One of his most eccentric theories was that life was carried to Earth — fully formed — by eggs in comets. But considering that he figured out how all the elements came to be (see page 248), we'll forgive him for that one.

In the 1940s, Fred turned to the universe. By that time, most scientists were happy that it began in a "cosmic egg" that exploded. Not Fred. He found the whole idea inelegant, and with his colleagues, mathematician Hermann Bondi and astrophysicist Tommy Gold, came up with an alternative. He even denigrated the "cosmic egg" theory, nicknaming it the Big Bang (which ironically made it *more* popular).

In 1948, the trio published their controversial theory of the universe: the Steady State. In their opinion, the cosmos had no beginning and it will have no end. This, of course, avoids the question about where it came from. But the breakaway faction had to answer how the universe can appear eternally changeless if the galaxies are speeding apart from each other.

Continuous creation, they concluded, is what fuels the cosmos. The formation of one hydrogen atom in a volume the size of a wine bottle every billion years would keep the universe topped up.

Cracks in the theory emerged almost instantly. The emerging science of radio astronomy revealed that distant galaxies were more tightly packed together — hardly a sign of an unchanging universe.

Fred and his colleagues stuck to their guns until the hammer-blow struck. The discovery of the cosmic microwave background in 1964 (see page 240) was the death knell for the Steady State. It could be nothing other than the afterglow of a hot Big Bang.

The theorists backing Fred were devastated. One of Heather's tutors at Oxford — Dennis Sciama — told her that he'd cried on hearing of the discovery. "The Steady State was such a beautiful theory," he said. At that point, he abandoned it.

Gold, Bondi and Hoyle getting up to mischief!

How will the universe end?

Astronomers know that our cosmos began in a Big Bang 13.8 billion years ago. But can we predict the future and say how it will end?

It's pretty certain that the universe is running down — and growing dimmer. The Big Bang created vast amounts of fresh gas that formed into galaxies made of blazing stars, like the Sun, often accompanied by planets like the Earth. As time went by, the stars died off, one by one. Now there's less and less gas around to form fresh generations of stars and planets.

In 2015, an international team of astronomers measured just how much the universe is fading. They found that our cosmos is only half as bright now as it was two billion years ago. We're on a one-way street to a cosmic blackout.

To make the picture even bleaker, all the galaxies are rushing away from each other. Astronomers once thought that gravity might put the brakes on this expansion, and pull everything together in a final incandescent smash-up called the Big Crunch. But they've recently found that a mysterious force (see page 235) is making the galaxies rush apart faster and faster. We can now predict that the future of our cosmos is to

become not only duller, but emptier, until all that's left is black dead stars in a vast cosmic wasteland. The words of the poet T.S. Eliot — written even before the Big Bang was discovered — are eerily prescient: "This is the way the world ends / Not with a bang but a whimper."

One day our universe will be dead as a dodo.

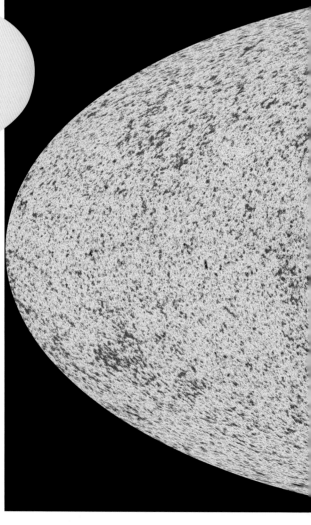

How cold is the universe?

It was pigeon poo that led astronomers to measure the temperature of the universe — and a lot more besides!

In 1964, physicists Arno Penzias and Bob Wilson were using a 6-meter horn antenna in Holmdel, New Jersey, to search for microwaves (short-wavelength radio waves) from our Milky Way's vast halo.

The minute they switched the dish on, they knew that something was amiss. It was picking up a signal, apparently coming uniformly from the whole sky. Was there something wrong with the telescope? Yes! It was covered in pigeon droppings. Penzias and Wilson cleaned up the mess and scared away the pigeons. But these homing pigeons came back — twice. Finally, the team deterred them "by decisive means."

Still the signal kept coming, 24/7. "We frankly didn't know what to do with our result," despaired Penzias.

The signal clearly wasn't coming from the Galaxy. But what about the whole universe? The team recalled that theoretical physicist Bob Dicke had been calculating how warm the cosmos would be now, billions of years after the inferno of the Big Bang.

Dicke came to look at the team's signal. It was everything he'd hoped for: the universe was bathed in radiation of 2.7 degrees Kelvin — that's −270°C (−455°F) — just a smidgen above the absolute zero of the temperature scale.

It's the outcome of the ultimate act of cosmic violence — radiation from the Big Bang cooled down by the expansion of the universe. And it's absolute proof that the Big Bang *did* take place.

Although it's officially called the cosmic

Seeds of creation: the Planck satellite reveals tiny temperature differences in the universe that will lead to the birth of the first galaxies.

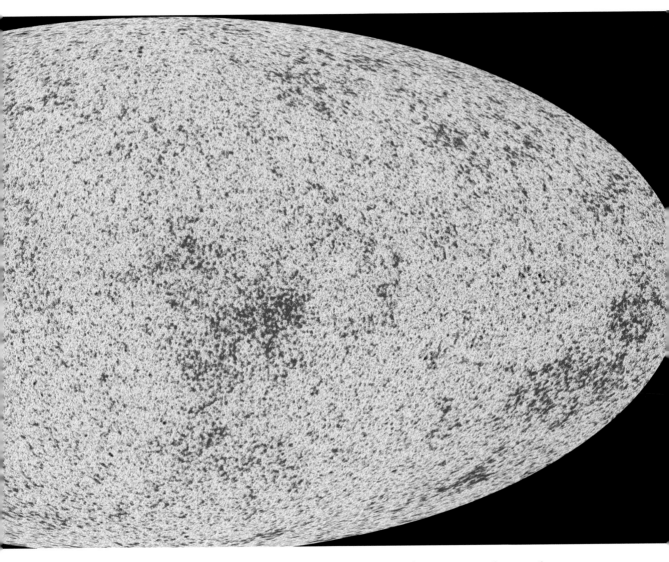

microwave background, astronomers refer to it as "the afterglow of creation." Immediately after the Big Bang, the universe was a frenzy of subatomic particles colliding with beams of light, making the environment an impenetrable white-hot fog. After 380,000 years, the temperature dropped low enough for electrons to bond with protons and make atoms. Light doesn't do much business with atoms, and that's when the cosmos became transparent. Radiation streamed outward across the universe as the microwave background.

The map imaged by Europe's Planck satellite shows tiny temperature fluctuations in the smooth fabric of the microwave background. Without these ripples in the primeval fireball, we wouldn't be here, because these are the embryos of the galaxies, stars and planets that surround us today.

TIME AND SPACE 13

Why do we have leap years and leap seconds?

Cleopatra got the whole thing going. She was taking a Nile cruise with Julius Caesar, who had just kicked her brother off the Egyptian throne, and they got talking dirty — about the calendar, that is.

The Earth takes 365 days and 6 hours to orbit the Sun, and if your calendar is fixed at 365 days, those odd hours add up remarkably quickly. Caesar was only too aware of the problem — back home in Rome, the calendar dates had surged forward so much that people were celebrating the March spring festival in the middle of winter.

Cleopatra's tame astronomer, Sosigenes, suggested that Caesar add an extra day once every four years, to bring the calendar back into sync. And that's what Julius did, making every fourth year a "leap year" by giving poor 28-day February an extra day.

It wasn't a perfect solution. That extra 6 hours is, more accurately, 5 hours and 49 minutes. As the centuries passed, the calendar slowly drifted again until, in 1582, Pope Gregory XIII tweaked the rules a bit. A century year would only be a leap year if you could divide it by four. So 2000 was a leap year, but not 1900 and 2100.

While leap years help fit days into the year, leap seconds involve slotting the right number of seconds into the day. The problem here is that the Earth's spin is slowing down, so the day is gradually growing longer. For instance, instead of being exactly 24 hours, the day we're writing this sentence is 24 hours 0 minutes 0.0008214 seconds long.

If we don't correct this drift, eventually we'd find the clock saying midnight when it's light outside. So, when the clocks drift away from the Earth's rotation by 0.9 seconds, we add a leap second — in effect, we stop our clocks for a second so that the sluggish Earth can catch up.

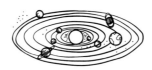

Julius Caesar gives Cleopatra the throne of Egypt; in return, she gives him leap years.

Are there parallel universes?

PARALLEL UNIVERSES - FIRST ASTRONAUT ON THE MOON

In the minds of scientists and mathematicians, yes. And they all seem to mean different things!

What parallel-universe theories have in common is that they describe universes existing somewhere we can't probe with telescopes.

One version is the "multiverse." If our Big Bang was a bubble breaking off from a pre-existing "quantum foam" (see page 231), then other bubbles most likely have split away too, to create other expanding cosmoses that aren't in touch with us.

But there's a much more interesting parallel-universe theory, which formed the plot of the movie *Sliding Doors.* A young woman tries to jump onto a train when the doors are closing. The film shows the two parallel lives that she then follows. In one, she makes the train; in the other, she is shut out — and they lead to totally different outcomes.

Some scientists think this happens for real. Suppose you put all your money on a spin of the roulette wheel: you either get very rich, or, more likely, very poor. But in the parallel-universe theory, both outcomes happen, and two different universes branch off from each other. One "you"

has to live as a pauper, while the other "you" has a superyacht and villas around the world.

It's not only down to our own decisions either. Every time there's a chance event, every possible outcome happens in one of many different universes that spring from it.

In our universe, for instance, the dinosaurs were wiped out by a space rock 66 million years ago. It left a niche for little mammals to evolve, and eventually evolve into humans. But the parallel-universe theory says that there's another universe — just as real — where that rock missed the Earth; the dinosaurs flourished while the mammals carried on hiding. In the end, it was dinosaurs who became intelligent and invented civilization, culture and the technology to explore the depths of space.

In a parallel universe, it was one small step for dinosaurkind . . .

How can we know so much about the universe when it's all so far away?

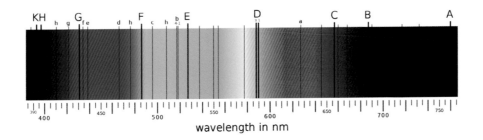

wavelength in nm

A dults have a terror of astronomical distances, and many regard the subject as an arcane art: "Well, it's all theory, isn't it?"

Emphatically, no! Even distinguished philosophers got it wrong. A century and a half ago, French philosopher Auguste Comte wrote: "The field of positive philosophy lies wholly within the limits of our Solar System . . . there is no means by which we will ever be able to examine the chemical composition of the stars."

Even before Comte wrote those words, a talented young German glassmaker, Joseph von Fraunhofer, had started the ball rolling. He made exquisitely accurate prisms and tested them on the Sun. He was perturbed to discover that its familiar rainbow spectrum — ranging in color from short-wavelength blue light to long-wavelength red — was crossed by dark, vertical lines. Being not a scientist but a superb technician, Fraunhofer mapped 570 of the lines, and got back to making prisms and lenses.

Fast-forward half a century to the University of Heidelberg. Chemist Robert Bunsen and physicist Gustav Kirchhoff were intrigued by Fraunhofer's discovery. When they heated different elements — including sodium, potassium and calcium — in the flame of a Bunsen burner and passed the resulting light through a prism, they saw different patterns of bright lines. Each element had its own characteristic spectral signature.

Bunsen and Kirchhoff realized these were the same lines that Fraunhofer had found in the Sun's spectrum. The team were now into cosmic forensic science — they could take the "fingerprints" of the Sun and the stars to work out their unique chemical compositions. In honor of the young prism-maker, celestial fingerprints were named "Fraunhofer lines."

Not only is the spectrum of a star a guide to its makeup, it also reveals its temperature, its age, and how fast it's moving. Spectroscopy is key to our understanding of the universe.

Key to the cosmos: the spectrum of the Sun, crossed with Fraunhofer lines

Where do the elements come from?

In the 1950s, the heated discussion in astrophysics was the origin of the universe: Big Bang or Steady State? A linked question was the creation of the elements. The Big Bang supporters, under brilliant physicist George Gamow, held that all the elements were created in the violent birth of the universe. Fred Hoyle and his Steady State allies turned to the stars instead.

When Fred began to work on the enterprise of "stellar nucleosynthesis," he joined forces with an international team of leading scientists: Margaret Burbidge, a superb observational astronomer, Geoffrey Burbidge, a gifted astrophysicist, and William Fowler, a Nobel Prize-winning nuclear physicist — the four were B²FH for short.

Astronomers already knew that stars are cosmic crucibles — in the fearsome heat of their cores, they fuse hydrogen into helium. But could they go further down the fusion route?

Fred realized that the next step had to be carbon, and suggested a new reaction that would convert three helium nuclei into one carbon

B²FH in Cambridge with Willy's 60th birthday present — a Mamod steam engine

The Origin of the Solar System Elements

Legend:
- big bang fusion
- cosmic ray fission
- merging neutron stars?
- exploding massive stars
- dying low mass stars
- exploding white dwarfs

1 H																	2 He
3 Li	4 Be											5 B	6 C	7 N	8 O	9 F	10 Ne
11 Na	12 Mg											13 Al	14 Si	15 P	16 S	17 Cl	18 Ar
19 K	20 Ca	21 Sc	22 Ti	23 V	24 Cr	25 Mn	26 Fe	27 Co	28 Ni	29 Cu	30 Zn	31 Ga	32 Ge	33 As	34 Se	35 Br	36 Kr
37 Rb	38 Sr	39 Y	40 Zr	41 Nb	42 Mo	43 Tc	44 Ru	45 Rh	46 Pd	47 Ag	48 Cd	49 In	50 Sn	51 Sb	52 Te	53 I	54 Xe
55 Cs	56 Ba		72 Hf	73 Ta	74 W	75 Re	76 Os	77 Ir	78 Pt	79 Au	80 Hg	81 Tl	82 Pb	83 Bi	84 Po	85 At	86 Rn
87 Fr	88 Ra																

57 La	58 Ce	59 Pr	60 Nd	61 Pm	62 Sm	63 Eu	64 Gd	65 Tb	66 Dy	67 Ho	68 Er	69 Tm	70 Yb	71 Lu
89 Ac	90 Th	91 Pa	92 U	93 Np	94 Pu	Very radioactive isotopes; nothing left from stars								

nucleus. For it to work, the carbon had to vibrate with a particular energy. At Fred's insistence, Willy examined it in his lab at Caltech, and found a carbon nucleus of the right resonance.

The way ahead was clear to build on the carbon breakthrough. A sufficiently massive star would build up successively heavier elements in its core: oxygen, neon, magnesium, silicon — all the elements up to iron. The iron core collapses, and a truly massive star eventually destroys itself by exploding as a supernova (see page 196). It vomits a cocktail of freshly minted elements into space.

Since then, astronomers have discovered that some of the heavy elements — including the gold in your wedding ring — aren't actually made inside stars. Instead, they're forged by exploding white dwarfs or the collision of neutron stars, which weren't even known in the 1950s (see page 252).

The source of all the elements, as we know it today, is shown brilliantly in the periodic table above, created by Ohio astronomer Jennifer A. Johnson. A few elements come from the Big Bang or are the debris from atomic collisions in space (cosmic ray fission), but as Fred suggested, the vast majority are made by stars.

The 1957 paper published by Fred's team — "Synthesis of the Elements in Stars" — first proved the overarching importance of the alchemy taking place within stars. One of the most important scientific papers of the 20th century, it's never referred to by its title. It's known affectionately in astronomy circles as "B²FH."

What causes gravity?

It was just the kind of chat you'd have over afternoon tea in the garden — if your companion happened to be Isaac Newton.

"Why should that apple always descend perpendicularly to the ground?" he mused to William Stukeley (the antiquarian who first investigated Stonehenge). "The reason is that the Earth draws it . . . and the apple draws the Earth." It was the first inkling of Newton's most famous discovery: the law of gravity. He went on to say that everything in the universe attracts everything else — accounting for why planets orbit stars and (after Newton's time) how galaxies rotate.

But Newton couldn't explain what gravity actually *was,* or how it could operate at a distance through the vacuum of space. The German mathematician Gottfried Wilhelm Leibniz even described Newton's theory as being both "unreasonable and occult."

In 1915, Albert Einstein devised a new theory of gravity: general relativity. Einstein said space wasn't really empty; the universe exists in a web of space-time. Okay, that's a bit mind-boggling. It's easier if we think in 2D — then space-time is like a thin rubber sheet. If you put a heavy weight on it, you'd make a dent. The resulting hollow in space-time is gravity.

Imagine the Sun in the center; the planets are orbiting it because they whirl around in this "gravitational well." Replace the Sun with a dense neutron star, and the walls of the dent become much steeper — it has stronger gravity. And if the object in the middle collapses completely to become a black hole, the dent becomes a funnel — travel to the center, and you fall out of our universe altogether.

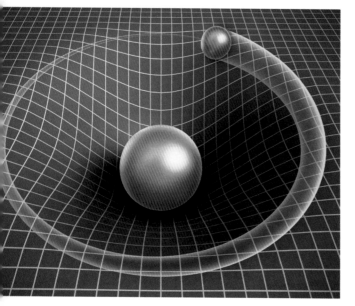

Einstein's way of looking at gravity: the Earth is trapped in the dent that the Sun (center) makes in space-time

What does E = mc² mean?

In 1905, Albert Einstein was pondering how a body's energy increases as it speeds up. We all know that a football has a lot more oomph when it's moving than when it's just sitting on the field. But his theory of special relativity told Einstein that even when it's stationary, a football , or anything else, contains energy.

To keep up its present luminosity, the Sun has to convert four million tons of matter into energy *every second*.

If you're looking at a ball with a mass — "m" — then you can calculate the energy — "E" — it contains from the formula $E = mc^2$, where "c" is the speed of light, and you have to square it.

The speed of light is a big number, and squaring it makes a *vast* number. That means a little bit of mass contains a huge amount of energy. If you could convert your football completely into energy, you'd have an explosion equal to detonating 10 million tons of TNT!

That's why nuclear bombs are so powerful. On a more peaceful note, it's what makes stars shine. In the Sun, the nuclei of four hydrogen atoms combine together to make one of helium. The helium nucleus is 0.7 percent lighter than four hydrogens, and the mass that's lost turns into energy, making the Sun shine. To keep up its present luminosity, the Sun has to convert four million tons of matter into energy *every second*.

The Ivy King atom-bomb test converted mass into energy on an awesome scale.

What is a gravitational wave?

On September 14, 2015, the world of science was shaken, both figuratively and literally. A gravitational wave from deep space passed through the Earth, and for the first time in history scientists captured its fleeting passage.

A century earlier, Albert Einstein had predicted these undulations in space-time. As we saw on page 250, Einstein visualized space-time as a rubber sheet. If you have two objects orbiting very rapidly, they'll send out ripples in space-time — just as you can create waves in the surface of a pond by stirring it with a stick.

The LIGO gravitational-wave telescope has two mirrors suspended at the end of a tube 4 km long; a laser beam is reflected backward and forward to measure if the mirrors are being swung back and forth by a passing ripple in space-time. The problem? A gravitational wave would move the mirrors by much less than the diameter of an atom — in fact, less than the size of a proton.

Still, by 2015 the system was up and running, and LIGO telescopes in Washington state and Louisiana simultaneously registered a shudder as a gravitational wave passed through. It was the first of several ripples that astronomers have now found coming from pairs of black holes in distant galaxies, emitting gravitational waves as they spiral together to form a single massive black hole.

August 2017 saw an even more exciting event. With the help of VIRGO, a newly commissioned gravitational-wave telescope in Italy, astronomers tracked down a gravitational wave to a much nearer galaxy.

Over a quarter of the world's astronomers swung into action, pointing telescopes on every continent — including Antarctica — toward the constellation Hydra to see what had happened. The telescopes picked out the exploding fireball from two neutron stars (see page 195) that had spiraled together and merged to make a black hole. The light from the explosion revealed that the eruption spewed newly minted precious metals into space — including 10 times the Earth's weight in gold!

Ripples in space-time: gravitational waves from a pair of merging black holes

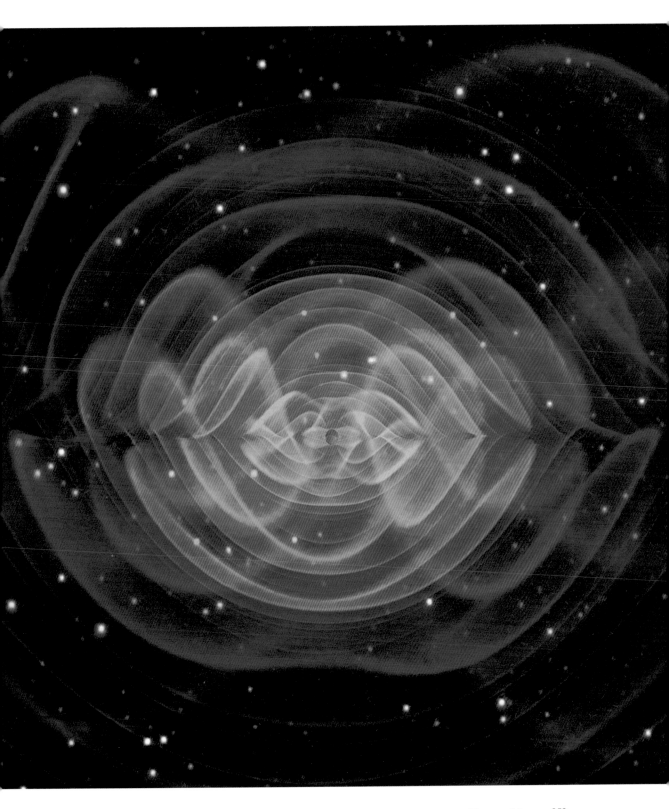

Is time travel possible?

Actually, it's easy — and you're doing it already! We are all time-traveling into the future, at the rate of one day every day. There are ways of speeding that up, if you want to visit the far future before you die. You could preserve your body cryogenically, to be thawed out centuries from now.

Perhaps the strongest evidence that it's impossible to travel to the past is that we don't have time-travelers from the future around us.

Einstein's theory of relativity gives you more advanced options: if you travel out into space fast enough, time slows down for you; when you return, you might be only a few months older, but thousands of years could have passed at home. The record-holding time-traveler is Russian cosmonaut Sergei Krikalev, who's spent so long in orbit that he is one-fiftieth of a second younger than if he had stayed on Earth.

But no one has yet traveled to the past. If we can ever create a wormhole (see page 208), it might provide a tunnel to our own past — but that technology is far beyond us at the moment.

And then there's the "grandmother paradox." Imagine you went back in time and accidentally killed your grandmother. Then she wouldn't have given birth to your mother, so you would not

A plutonium-powered DeLorean is as plausible as any other time machine.

have been born, so you couldn't go back and kill her.

The classic movie *Back to the Future* explores this paradox in a more entertaining way: Marty accidentally ends up in the past, courtesy of a time-traveling DeLorean car, and his mother — then a teenager — falls in love with him. It takes some quick thinking on Marty's part to get his parents together so that he will actually exist.

Perhaps the strongest evidence that it's impossible to travel to the past is that we don't have time-travelers from the future around us. If our distant descendants could travel freely in time, tourist trips to the past would surely be highly popular!

ALIEN LIFE 14

Is there anybody out there?

Of all the questions we get asked, this has to be IT. Is there life "off Earth"? What is it? What does it look like?

Life *on* Earth shows a huge range of diversity: from bees to bison, mushrooms to mice, humans to humpback whales. But what it all has in common is that living beings are made up of carbon and water — two of the most widespread materials in the universe.

So, if here, why not there? Astronomers have tentative evidence for life on Mars, and researchers believe that below the surface of Saturn's moon Enceladus there could be a deep ocean harboring cosmic crustaceans. But these are hardly advanced forms of life, and the consensus among astronomers is that, if there *is* life out there, it's more likely to be little green slime, rather than little green men. However, life evolved on our own planet from slime to intelligence in a short time on the cosmic timescale — in just a few billion years (the universe has existed for 13.8 billion years). So, might bacteria have become mega-brains on other worlds?

In the last twenty years, astronomers have discovered some 4,000 planets circling other stars. Could they be abodes for intelligent life?

Researchers are on the job. Over 60 years ago, radio astronomer Frank Drake wondered if intelligent aliens could be broadcasting to us — like a powerful cosmic radio station — to send information about themselves.

Drake and his colleagues set up an initiative called SETI — the Search for Extraterrestrial Intelligence. They've used ever more powerful radio telescopes to tune in to the cosmos — the latest and most sensitive ear on the universe being the Allen Telescope Array, funded by Paul Allen, cofounder of Microsoft.

The investigation hasn't yet come up with a whisper of a signal, but there's still hope. As far as the universe is concerned, we Earthlings are the new kids on the block. Perhaps radio transmission as a means to communicate across the cosmos is hopelessly outdated. As one SETI researcher, Dan Werthimer, explained:

"We're also looking for laser signals. If you'd asked me 200 years ago how we might communicate with extraterrestrial civilizations, I might have said that smoke signals are the best way. If you ask me 200 years from now, there might be something even better — perhaps faster-than-light particles called tachyons, or some sort of new particle we don't even know about."

Frank Drake: father of SETI — the Search for Extraterrestrial Intelligence

How did life on Earth begin?

Charles Darwin was a man way ahead of his time.

In 1871, he wrote to a friend about how life might have begun on Earth: "if (& oh what a big if) we could conceive in some warm little pond with all sorts of ammonia & phosphoric salts, — light, heat, electricity &c present, that a protein compound was chemically formed . . ."

Darwin was exceptionally close to today's thinking.

It certainly happened quickly. The Earth was born around 4.6 billion years ago; the first primitive cells appeared less than a billion years later, 3.8 billion years in the past. The implications for life in the universe — and its subsequent evolution — are huge. Since life arose rapidly on Earth, the same might apply to the thousands of other worlds that astronomers are discovering in our Galaxy.

So where did the raw materials of life come from? Some scientists say that our traditional cosmic foes, comets and asteroids — well known for causing devastation — may also have been our friends, delivering to early Earth the water and carbon-rich compounds that made the essential building blocks of life.

Darwin's "warm little pond" has seen resurgence in popularity recently. Biologists believe that ancient rock pools, alternately filling and drying up, could have concentrated primitive compounds of carbon, hydrogen, oxygen and nitrogen (incidentally, the most common elements in the universe) into more complex molecular entities.

Since life arose rapidly on Earth, the same might apply to the thousands of other worlds that astronomers are discovering in our Galaxy.

Key to the processes of life is the giant molecule RNA. A precursor to the more complex DNA, upon which all life today depends, and which governs the behavior of our 50 billion cells — it was probably the first chemical compound to self-replicate. By copying itself over and over again, RNA could build up complex structures that could then replicate themselves. It was the beginning of life; a process that may have taken just 200 to 300 million years.

But there may be more durable sites for life's origins than fragile rock pools. Our deep oceans are riddled with mid-ocean ridges, where the Earth's internal heat is jostling its vast tectonic plates.

Enter hydrothermal vents — volcanic fissures where superheated water erupts into the cold ocean depths. Hottest of all are the "black smokers," with temperatures of over 400°C (750°F). Discovered by the ALVIN submersible in 1979, black smokers are a paradise for marine life. With no sunlight to provide energy, giant tube-worms, shrimps and snails feed on bacteria that thrive on the smokers' chemical energy.

That same chemical energy may have kicked off life, many scientists now believe. It brewed up the mighty molecules of life — like RNA — within tiny pores in the rock. Once this crucible of life could surround itself with a membrane, it floated away as a living cell.

Similar life-giving vents may exist on the ocean floors of other worlds in our Solar System. The likeliest candidates are Jupiter's moon Europa and Saturn's moon Enceladus, and missions to these distant destinations are on the drawing board.

Crucible of creation? A black smoker deep in the Pacific Ocean teems with life.

What are the most extreme conditions life can survive?

This cute little critter, just 1 mm long, has secrets of life that we mere mammals should know about.

Tardigrades — of which there are over a thousand species — are the closest we get on Earth to indestructible. They can survive just about anything that life can throw at them, provided they have enough water.

Let's go to the extremes these cuddly micro-animals can cope with: heights (mountains in the Himalayas over 5 km high), extreme pressures that would flatten other animal life, hot springs, deep-sea vents, temperatures ranging from just above absolute zero to over 150°C (300°F) . . . and that's just for starters.

These water bears have been with us for 500 million years. You can see them with a kid's microscope, creeping through moss, their favorite habitat (their name means slow-stepping). People get so attached to them that they even buy tardigrade toys!

Tardigrades are built for defense. Their eight legs end in ferocious claws; their teeth are like daggers. And they have a phenomenal survival mechanism: when the going gets hot, and water dries up, they dry up too, and can exist for decades in suspended animation.

Bring back the water, and the tardigrades wake up, eating antioxidants like vitamin C to repair any damage to the DNA in their cells.

But their most impressive survival feat took place in 2007, when a colony of tardigrades was blasted into the vacuum of space — an environment that would destroy a human being (see page 60). Not only did most of them survive the vacuum, they coped with the lethal radiation; some females even laid eggs in space, giving rise to healthy offspring back on Earth.

What's their trick? Being able to patch up their DNA — the roadmap for their cells — is part of it. But biologists are watching them closely to pick up more clues as to how *we* can survive better.

If and when the next mass extinction comes along, we won't make it through, but the tardigrades will.

Survival of the fittest — the tardigrade (water bear)

Is there evidence aliens have landed on Earth?

This question has its place in the ranks of classic conspiracy theories, and it is certainly a biggie. It even has self-proclaimed "expert ufologists" arguing among themselves . . .

Let's start with the least contentious question first. What are the Nazca Lines, which crisscross the Pampa Colorada plain in Peru? These vast lines, the largest of which is 370 m long, were created by the Nazca civilization (500 BC–AD 500) by removing red pebbles from the desert's surface and exposing the gray soil underneath. Spectacular in their scale, they portray birds, animals, trees, fish and humans. The hummingbird is 93 m across; the spider, 47 m!

An image of this mummified Native American child was presented as proof of alien life, and later debunked as a hoax.

Historians originally contemplated that the lines might be an ancient observatory, like Chankillo (also in Peru), whose stone towers chart the progress of the Sun along the horizon throughout the year, and predict the seasons. That idea didn't pass muster, nor did the idea that the lines were constellations mapped onto the desert. There's no doubt, however, that they were created by people.

But it's amazing how you can influence people if you're a ufologist. A certain discredited "UFO expert" published a book entitled *Chariots of the Gods*. To him, monuments like the Nazca Lines, Stonehenge and the pyramids were created by gods in flying saucers to help them navigate their craft back to Earth, and many people believed him.

It can only get worse. In early July 1947, something crash-landed in the remote countryside near Roswell, New Mexico. It gave rise to "the world's most famous, most exhaustively investigated, and most thoroughly debunked UFO claim," in the words of one investigator.

The U.S. Air Force admitted that the "something" was a helium balloon, then promptly shut up, prompting conspiracy theories. They escalated spectacularly 30 years later, when rival teams

of ufologists began to publish ever more embellished accounts — it was a great little business!

The USAF clarified matters in the 1990s, when they admitted that the balloon had been on a nuclear test surveillance mission for their secret Project Mogul. But it was too late. Rumors of four alien bodies discovered at the crash site were now circulating.

In 2012, a supposed image of one of the aliens came to light. It had been taken by the then-deceased geologist Bernard Ray and his wife, Hilda, and had been kept in a box for 50 years.

When it was finally released, it caused alien furor. Thousands packed halls and convention centers across the U.S. to find out the truth. Unfortunately for the ufologists, a label that's visible in some of these photographs reads "MUMMIFIED BODY OF TWO-YEAR-OLD BOY." The alien proved to be the remains of a Native American child found in 1896, which was once on display at the Chapin Mesa Archaeological Museum in Mesa Verde, Colorado. To our knowledge, it's still in storage at the museum.

The Nazca lines in Peru were created by ancient humans — not by aliens.

Are there canals on Mars?

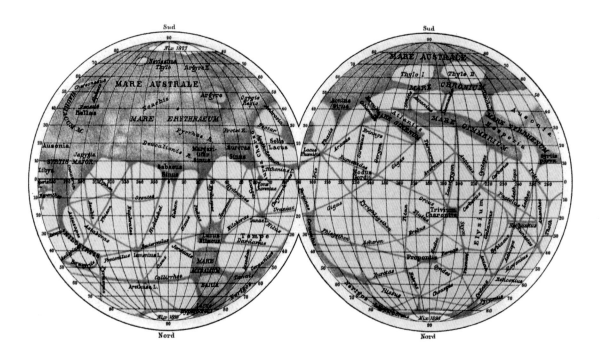

Every two years, Mars swims closer to the Earth than usual. On one of those occasions, in 1877, the Italian astronomer Giovanni Schiaparelli observed regular lines crossing the planet's surface. He called them *canali* — channels.

Word spread quickly to the States, and to a rich Boston businessman and amateur astronomer, Percival Lowell. His grasp of Italian was clearly not great, and he became convinced that these linear features were artificial canals built by Martians to save their dying world from drying out.

Dedicated to finding life on Mars, Lowell even built an ambitious observatory 2,210 m up in the mountains above Flagstaff, Arizona. He created wonderful globes of Mars depicting in amazing detail the canal network he was observing. And Lowell spent much of his life writing about Martians and their superior intelligence, spawning some sensational sci-fi epics.

But in the end, the canals turned out to be an illusion, an artifact of the optics of old-fashioned lens telescopes and the vagaries of the human eye. But their discovery and Lowell's popularization of them led to a widespread and persistent belief in advanced life on Mars.

1888 Atlas of Mars by Giovanni Schiaparelli. This telescope view shows south at the top.

What's the Face on Mars?

R eady for another conspiracy theory? Mars seems to attract them like flies around a trashcan. And everyone, it seems, has heard of the Face on Mars.

In the summer of 1976, NASA's *Viking 1* orbiter imaged Cydonia, a flat plain on Mars. The low-resolution view revealed a jumble of features, one of which looked uncannily like a dark-eyed man wearing a helmet. It measured 1.6 km across and was over 400 m high.

Richard Hoagland, an unqualified former museum curator, latched on to it with fervor. It was proof, he declared, that Mars had, or once had, intelligent life. What else, he argued, could have created a monument as vast as this; even mightier than the pyramids?

Hoagland was evangelical about "The Face,"

and attracted an army of disciples. We were unfortunate enough to bump into some of them, protesting outside the gates of NASA's Jet Propulsion Laboratory, blaming the space agency for blowing up its *Mars Observer* probe to avoid admitting that there was life on the Red Planet.

Fast-forward to 1998. NASA was poised to launch its next Mars probe, the *Mars Global Surveyor.* Mike Malin, the world's most brilliant planetary imager, was charged with photographing the Face. He railed against NASA. "They told me to sacrifice science for the Face," he told us.

With his much higher-resolution camera, Malin reluctantly delivered the goods. And the Face appears not so much sculptural as geological. Malin's colleague Ken Edgett points out, "It looks like a big butte in Monument Valley."

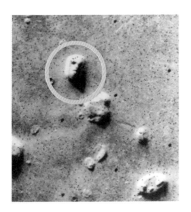 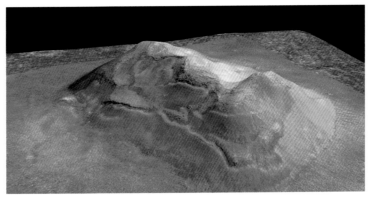

Now you see it, now you don't: the Face on Mars turns out to be an eroded mesa.

Is there life on Mars?

In the 20th century, even sceptical astronomers were prepared to go some way toward admitting that there could be *primitive* life on Mars. As kids, we learned that the dark markings on Mars were vegetation that ebbed and flowed with the seasons. They weren't. The markings turned out to be patches periodically covered and uncovered by wind-borne dust.

The real test for life on Mars came with the arrival of NASA's twin *Viking* probes in 1976. Each contained four experiments designed to find life in the form of "green slime."

One of the experiments was designed by Gil Levin; not a regular NASA scientist, but a highly successful sanitary engineer — an expert in finding Legionnaires' disease bugs in air-conditioning systems.

The "Labeled Release" experiment was simplicity itself: put a Martian soil sample in water, feed it nutrients spiced with a radioactive marker, and wait. If there are any bugs present, they'll eat the nutrient, burp and release radioactive gas. As the bugs reproduced, the gas would build up until the nutrient had all been consumed.

And that's what Levin's experiment revealed. He showed us the results in his office off the Beltway in Washington — his curves for Mars exactly matched the data he had collected for bacteria on Earth.

Geologists have ample evidence that water once flooded Mars; now it undoubtedly exists in the ground as permafrost.

Water on Mars? Osuga Valles appear to have been carved out by intense floods in the past. Image from Europe's *Mars Express* space probe.

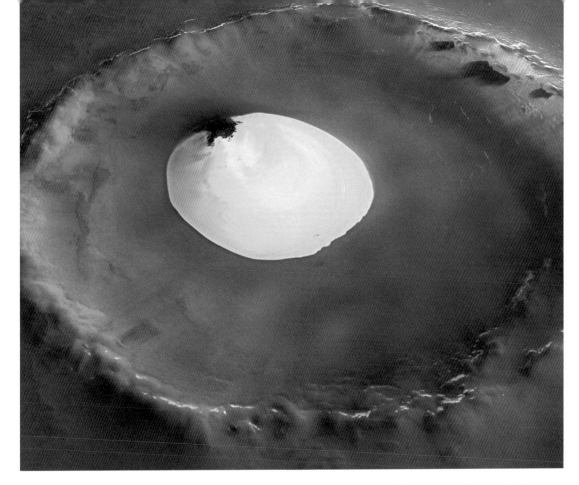

But another experiment — NASA's gas chromatograph–mass spectrometer (GCMS) — had found no carbon in the Martian soil. Without this essential element, there couldn't be living organisms. NASA had to come up with a definite answer to the question of life on Mars, and they chose "No."

Much later, a new wave of scientists pointed out that the GCMS was so grossly insensitive that it would have missed out on 30 million bacteria cells per gram of soil. So the case for life on Mars is, in fact, still wide open.

The problem is that post-*Viking* NASA hasn't had a biologist leading a mission to Mars. Other scientists have had a field day peering down from orbiters and trundling rovers all over the Red Planet — and they keep coming up with evidence that Mars could be biologically active.

Take water, the lubricant of life. Geologists have ample evidence (pictured above) that water once flooded Mars; now it undoubtedly exists in the ground as permafrost. In addition, meteorologists have found reactive methane in the Martian atmosphere. Is it belched out by dormant volcanoes, or is it emitted by living cells?

Now, the search for Martian life is on again. NASA's Mars 2020 mission will ferry a rover designed to drill for biological samples on the Red Planet — with a view to a future human mission (see Chapter 3).

A lake of frozen water 10 km across nestles within a crater near Mars's north pole.

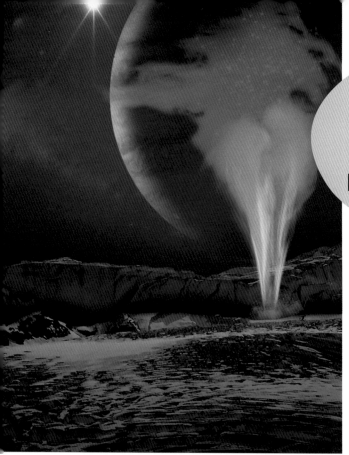

> ## Could there be life on other planets' moons?

Moons of the outer planets as abodes of life? It's possible.

Europa, the least colorful of Jupiter's major moons, is pure white and smoother than a billiard ball. That's because its surface is made of ice, and underneath there may lie a deep ocean.

The Hubble Space Telescope has spotted plumes of water erupting from Europa's south pole — more evidence of an ocean. As a result of gravitational pummeling by Jupiter, there are probably black smokers on the ocean floor (see page 261). It's an ideal environment for life.

Move out to Saturn, and you'll find Enceladus, one of the planet's many icy moons. Just 500 km across, this tiny world, like Europa, is geologi-cally active and could also harbor a deep ocean. Cryovolcanoes at its south pole eject jets of salty water laced with organic molecules, which means that Enceladus does contain at least the building blocks of life.

The strongest contender for extraterrestrial life is Saturn's giant moon, Titan. Larger than the planet Mercury, Titan boasts a thick atmosphere of nitrogen, with a pressure 50 percent greater than the Earth's. In 2005, Europe's *Huygens* probe landed on the moon's surface, reveal-ing a landscape of icy pebbles under a cloudy orange sky. Its mother craft, the orbiting *Cassini* probe, used radar to penetrate these clouds and discovered huge lakes of ethane and methane on Titan's frozen surface. These complex organic compounds are ideal precursors for life — all they need is a little heat to get them going.

The strongest contender for extraterrestrial life is Saturn's giant moon, Titan.

Artist's rendering of a plume of water vapor ejected by Jupiter's moon Europa

What are the most extreme exoplanets?

When astronomers began to search for extrasolar planets, they expected to find twins of the planets of the Solar System. Instead, they've turned up nightmare worlds.

The first exoplanet discovered, Dimidium (see next question), was a huge surprise: a planet bigger than Jupiter, but far closer to its parent star than Mercury is to our Sun. It's not the most extreme "hot Jupiter" though; that title is currently held by KELT-9b — so close to its brilliant sun that its dayside broils at a temperature of 4,300°C (7,800°F), hotter than the star Betelgeuse. This gaseous world is actually boiling away into space.

That fate has already befallen poor little Kepler-70b. Once it was a huge, hot Jupiter, but its star's heat has stripped away the planet's gases. All that's left is the planet's core, forming an incandescent rocky world smaller than the Earth.

Another hellish place is Poltergeist. It's a cool, rocky world a bit bigger than the Earth. The planet gets its spooky name from the fact that it's orbiting a pulsar — the spinning collapsed remains of a dead star (see page 195). This pulsar blasts the barren surface of Poltergeist with deadly beams of radiation (see the artwork above).

At the other end of the scale, OGLE-2005-BLG-390L lies so far from its star's warming glow that it is perpetually frozen at a temperature below −220°C (−360°F). The coldest planet known, its surface must be coated with ices made of substances that are gases on Earth — methane, ammonia and nitrogen.

Finally, there's PSO J318. To look at, it's a lot like Jupiter. But there's one big difference: this "rogue planet" does not have a sun. It was probably born circling a star, but the gravity of its sibling worlds tossed PSO J318 far out into space. Astronomers believe there are billions of giant rogue planets in the Milky Way, along with many more smaller ones. There may be more free-floating planets in our Galaxy than there are stars!

When your sun is a pulsar . . . expect to be zapped by radiation.

How do we find planets of other stars?

"The space we declare to be infinite . . . in it are an infinity of worlds of the same kind as our own."
— Giordano Bruno, 1584

Leading in the search for extrasolar planets was Dutch astronomer and talented composer Peter van de Kamp, who knew that the gravitational pull of a planet on its parent star would make it wobble by a tiny amount. Sadly, for most of the 20th century, the equipment wasn't able to measure these wobbles with any accuracy.

By the 1990s, modern technology was ready for the task. Everyone was prepared to be patient. After all, our biggest planet, Jupiter, which circles the Sun in about 12 years, causes a long-term wobble in our star's position on the same timescale. So in 1995, Swiss astronomers Michel Mayor and Didier Queloz were dumbfounded when the Sunlike star 51 Pegasi wobbled with a period of just *four days.* It pointed to a "hot Jupiter" — now called Dimidium — orbiting its star very closely (see previous question).

Other astronomers were quickly on the short-wobble trail. A team led by Geoff Marcy discovered 70 of the first 100 exoplanets, many of them in planetary systems.

Then a new technique came to hold center stage. Graduate student David Charbonneau (now a professor at Harvard) invented a way of finding an extrasolar planet by watching the dip in its star's light when the world crossed in front of it — all with a 10-cm telescope!

Charbonneau's vision led to the orbiting Kepler Space Telescope, which has discovered over 2,500 exoplanets. Now (March 2018) the grand

One of the Earthlike worlds circling the star TRAPPIST-1, seen here in an artist's impression, is among thousands of planets circling other suns.

total of our neighbor worlds stands at 3,743.

The statistics make for astonishing reading. One in five Sunlike stars has a planet in the "Goldilocks zone" where it's not too hot and not too cold for water — the prerequisite for life — to exist in liquid form. That means that if there are 200 billion stars in the Milky Way, there's a potential for 10 billion habitable planets. This rises to 40 billion if you include the ubiquitous red dwarfs (see Chapter 9); many have planetary systems, and a red dwarf's longevity enhances the possibility for life to evolve.

What we've learned about our extrasolar neighbors is amazing: their colors (red, blue and magenta), circling moons and rings, magnetic fields, volcanism, even atmospheres. But we still haven't discovered what we set out to find — another Earth with intelligent life.

If aliens exist, do they look like us?

Everyone knows what aliens look like. They have domed heads to contain their superbrains, bulging eyes to take in the panorama of the universe (from the best UFO seat), and they're bipedal (helps when you're using human actors).

But *please* could movie and TV makers be a bit more inventive with their science? If they knew the tiniest bit about physics and biology, they could create much better aliens.

To make a well-functioning alien, water is key. Whatever aliens are made of, they need a solvent to process the chemical reactions that power life. The rest is largely determined by gravity.

Imagine a high-gravity planet with a thick, murky atmosphere. Its inhabitants would be squat and possibly blind. They'd be in no hurry to do anything. You could be looking at a world of giant slugs, equipped with sonar to help them find their way around.

At the opposite extreme is a low-gravity planet. Aliens here could delight in growing tall, with dextrous limbs that would enable them to build tools and develop technology. Low-gravity aliens are likely to be a spacefaring species.

Thin atmosphere? Not a problem. The aliens could breathe through large, branching gills that are separate from the eating region. Plus, the thin air's transparency would give them a window to the universe — an incentive to explore space.

Sex and reproduction don't have to be internal. Tentacles would do just as well, and a low-gravity alien would be blessed with many to provide stability on its world (sorry, bipedal actors). External gestation would solve any problems related to squeezing a baby out from an internal womb through a narrow gap in the pelvis.

Vision would be essential, and tuned to the medium frequencies of light rather than radio waves or X-rays. That's the kind of radiation that stars produce most copiously.

And finally, food is anybody's guess. But we hope that, like us, they'd be veggie.

An exotic alien we created for a TV series. In the real universe, they'll probably look even more bizarre.

Carbon is the most user-friendly of all the elements. It loves bonding with other atoms, building up complex structures essential for underpinning the diversity of living organisms. In fact, carbon compounds are described as "organic" — pertaining to living things.

When we search for life in space, we focus on the essentials: water and carbon. We know plenty of places in the Solar System where water exists: Mars, the satellites of Jupiter and Saturn, and even our own Moon — but what about carbon? Is it *truly* essential?

Sci-fi writers, in particular, have come up with an alternative — silicon. Admittedly, it's not as cuddly as carbon, but it's still on friendly terms with other atoms. It's just that the bonds that silicon forms aren't as flexible as the links that carbon makes.

Silicon favors making crystalline bonds — like a collection of microchips. It's no coincidence that the technological heart of California is called Silicon Valley! Writers have speculated that cities of silicon microchips on asteroids could organize themselves into intelligent societies, not even needing water or an atmosphere. Trouble is that you'd have to be a top computer nerd to converse with them.

One of the most innovative suggestions for a *very* different form of alien life came from top astrophysicist Fred Hoyle — wearing his novelist's hat this time. In *The Black Cloud,* Fred imagines a dark cloud (like those in which stars are born — see page 186) that drifts into the Solar System. It's feeding on energy from the Sun, and it is intelligent, as the myriad molecules within the cloud communicate with each other like nerve cells in a brain.

> When we search for life in space, we focus on the essentials: water and carbon.

In the novel, the repercussions for life on Earth aren't pleasant, but scientists — including fictional Professor Chris Kingsley — realize that the complex cloud and its networks could tell them a thing or two about the universe.

Life . . . but not as we know it

Can I pick up alien signals at home?

Yes indeed. This picture is a screenshot of the SETI@home program operating on Nigel's laptop. It displays radio signals from space and works great as a screensaver too!

More than five million people are involved in the SETI@home project, tracking signals received by the Arecibo and Green Bank radio telescopes that may have been broadcast to us by extraterrestrials. The University of California, Berkeley, organized this "distributed computing" system that would outdo any supercomputer. They harness the global computing power of enthusiasts to analyze the mass of radio telescope signals they've collected.

As the signal shifts across your screen, you're scrolling the radio wavelengths that the telescopes have scanned. Look for a tall, narrow-wavelength spike — it may have been generated deliberately by an extraterrestrial civilization. Or it may just be radio-frequency interference . . . as of 2018, there are still no positive detections.

Now Berkeley has redoubled its efforts in tracking down aliens. In 2016, it started the Breakthrough Listen project. Funded to the tune of $100 million by philanthropist Yuri Milner, this is the mother of all SETI tools.

It collects data from three radio telescopes:

the 100-m at Green Bank, West Virginia; the 64-m dish at Parkes, New South Wales; and the newly commissioned FAST telescope in China, which, at 500 m across, is the biggest radio telescope in the world (see page 84).

These are joined by the 2.4-m Automated Planet Finder at California's Lick Observatory — an optical telescope that searches for laser beams directed from extraterrestrial communities.

This ambitious project will last for 10 years. It will target a million stars, and the centers of a hundred galaxies. Everyone can join in — just access Berkeley's Open Data Archive, where you can download the results for software analysis.

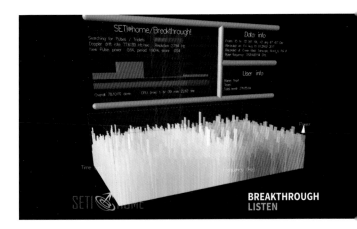

SETI@home — pick up the call from ET in your computer's downtime.

Our first message in a bottle to the stars didn't go down very well. It was posted on plaques aboard the twin space probes *Pioneers 10* and *11*, launched in the early 1970s toward Jupiter and Saturn. Scientists knew that both spacecraft would go interstellar — so why not tell our alien buddies out there a little about ourselves?

The *Pioneer* pictograms showed Earth within the Solar System, the Solar System's position in the Galaxy relative to the nearest pulsars (see page 195), and the composition of hydrogen. And a drawing of the beings who had sent it — naked!

Feminist critics objected that the woman wasn't raising her hand in greeting as the man did; middle-America was appalled that the plaque showed the couple's naughty bits. One enraged letter writer ranted that "NASA is using taxpayers' funds to send smut into space."

Rethink was required. On the next launch to the gas giant worlds — *Voyagers 1* and 2 in 1977 — NASA was careful not to reoffend. Instead, it bizarrely attached a golden phonograph record encoded with the sounds and images of Earth (see page 66).

In 1974, between the plaque and the record, an ace idea was hatched. The dish of the giant radio telescope at Arecibo, Puerto Rico, was resurfaced, and the director, Frank Drake (founder of SETI), wondered how to celebrate. His personal assistant suggested using the world's then-largest radio telescope in reverse to beam a message to our alien friends.

This time, the scientists devised a code for their message that would be easy for the aliens to crack. It was aimed at M13, a globular cluster 25,000 light years away — which means that we won't get a reply for 50,000 years!

The message consisted of 1,679 on–off pulses; the product of two

The Arecibo message: Will the aliens understand this message from Earth when post-grad students couldn't?

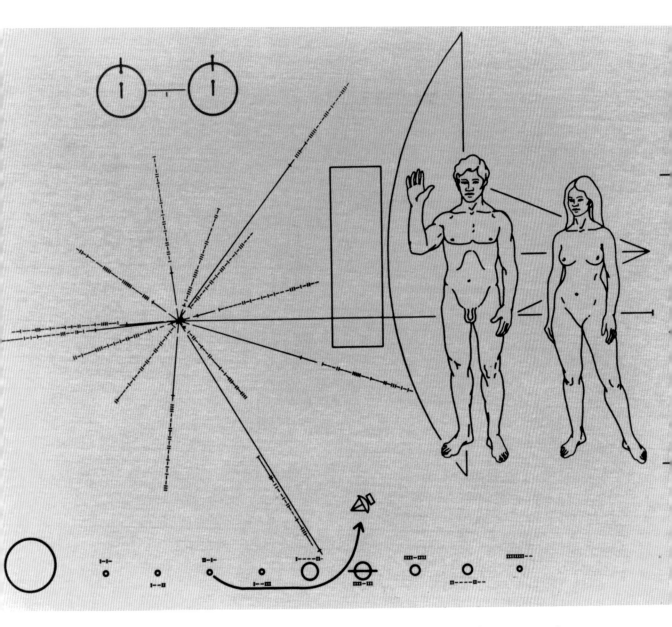

prime numbers — 23 and 73. An intelligent alien would know what to do next: arrange them into a rectangle and look at the pictogram.

Have a go. Before you try, you're warned that even science graduate students couldn't grasp it. But here are some hints: the first 10 blips are the vocabulary of the binary code; next come the essential molecules for life; the curly bits represent the structure of DNA, the backbone of life; then there are images of a human being, the Solar System and the telescope that sent the message.

Two identical plaques attached to *Pioneers 10* and *11* are heading for interstellar realms.

Should we reply?

What if we do pick up a signal from an alien civilization? Should we respond? Doom-mongers would have us believe that to "pick up the phone" would lead to an invasion of aliens hell-bent on destroying life on Earth.

They're dummies. We're the new kids on the block, cosmically speaking. Aliens are likely to be far more advanced than primitive human beings. To have survived so long, they will have sorted out the problems of war, famine and conflict.

The SETI community has its own international protocols in place to respond to an alien message that would involve the whole global community, not just the SETI scientists.

And it's too late anyway. Any aliens living within 100 light years of Earth will know that we exist. Since the 1920s, we've been churning out radio and TV programs with increasing signal strength. These broadcasts leak into the cosmos at the speed of light.

So, they know we're here, and they may be on their way.

Let's meet and greet. We'll have a lot to learn from them.

The Parkes radio telescope in Australia: about to open a radio dialogue with extraterrestrial life?

GLOSSARY

Accretion disk: A hot, bright disk of gas that circles a black hole.

Active galaxy: A galaxy whose center is experiencing violent activity, powered either by a burst of star formation or by a supermassive black hole.

Astronomical unit (AU): The distance from the Earth to the Sun (149.6 million kilometers). An AU is the standard unit used to measure distances in the Solar System

Asteroid: A small, often irregularly shaped Solar System object, made of rock and/or metal, that orbits the Sun generally between the orbits of Mars and Jupiter.

Big Bang: The explosion that created the Universe 13.8 billion years ago. From this extremely dense and hot state the cosmos has been expanding ever since.

Big Crunch: A now discredited theory proposing that the whole Universe will eventually collapse to a single point.

Binary star: Two stars in orbit around each other, under the influence of each other's gravity. Two-thirds of stars are in double-star systems.

Black hole: An object in space with such a strong gravitational pull that nothing, not even light, can escape from it.

Blue Moon: Originally a Moon colored blue by dust or smoke in Earth's atmosphere; now commonly used to mean the second Full Moon in a month.

Carbonaceous chondrite: A carbon-rich meteorite that originates from an asteroid's surface. They may have delivered the raw materials of life to the infant Earth.

Catadioptric telescope: A telescope which focuses light using both a mirror and a lens.

Coma: The glowing 'head' of a comet, which may become as big as the Sun, formed from gases boiling off the comet's icy nucleus.

Comet: A small icy object from the outer Solar System that can grow a gaseous coma if it approaches the Sun. Some comets also sprout tails, made of gas and dust, that may be millions of kilometres long.

Corona: The Sun's outer atmosphere, which can be seen during a total solar eclipse.

Coronal mass ejection (CME): A solar storm. CMEs eject dangerous streams of particles and radiation into our Solar System.

Cryovolcano: A cold volcano that ejects jets of icy water.

Dark energy: The energy that scientists believe is responsible for the accelerating expansion of the universe.

Earthshine: Sunlight reflected from the Earth onto the Moon, illuminating the dark portion of a crescent Moon.

Event horizon: The perimeter of a black hole, marking the point of no return for either matter or light falling in.

Exoplanet: A planet that orbits a star other than the Sun.

Galactic year: The period it takes our Sun to circle the Galaxy — roughly 225 million years.

Galaxy: A collection of millions or trillions of stars, gas, and dust held together by gravity.

Galaxy cluster: A swarm of hundreds or thousands of galaxies.

Galaxy group: A small collection of a few dozen galaxies, like the Local Group containing the Milky Way and Andromeda Galaxy.

General relativity: Einstein's theory of gravity.

Globular cluster: A densely packed cluster of up to a million very old stars.

Goldilocks zone: An area in a solar system where it's not too hot nor too cold for water to exist in liquid form.

Gravitational wave: The ripples in space-time created by two celestial objects orbiting each other very quickly.

Harvest Moon: The Full Moon that appears closest to the autumn equinox.

Infrared telescope: A telescope that picks up wavelengths between radio waves and light, emitted by warm — often hidden — objects in the cosmos, such as very young stars.

Kuiper belt: A region populated by small icy worlds — Kuiper belt objects — lying beyond the orbit of Neptune in the outer Solar System.

Light year: The distance traveled by light in a vacuum in one year; used to measure distances to stars and galaxies.

Lunar eclipse: When a Full Moon passes into the Earth's shadow and sunlight is cut off.

Meteor: A streak of light in the sky, also called a shooting star; seen when a meteoroid burns up on entering Earth's atmosphere.

Meteorite: A meteoroid that reaches the ground and survives impact.

Meteoroid: A particle of rock, metal, or ice traveling through space

Nebula: A cloud of gas and dust in space.

NEOs: Near-Earth Objects, asteroids or old comets that have the potential to hit Earth and cause destruction.

Neutrinos: Lightweight subatomic particles without electric charge, created in the core of stars and responsible for the explosion of a supernova.

Neutron star: See *pulsar.*

Nova: An explosion on the surface of a white dwarf when it accumulates gas from its companion star.

Parallax shift: A technique of measuring a star's distance by observing its apparent change in position when seen from opposite sides of the Earth's orbit.

Planet: A large object in orbit around a star.

Planetary nebula: A glowing cloud of gas — a star's former atmosphere ejected at the end of its life.

Planetesimals: Rocky and icy chunks of debris around a forming star that later amalgamate to create the planets.

Pulsar: A very dense, tiny star containing more matter than the Sun and mainly made up of subatomic particles called neutrons — hence also called a neutron star. As a pulsar rotates, radio waves from its poles sweep past the Earth as regular pulses.

Quasar: The bright disc of gas that circles a supermassive black hole in the middle of a galaxy.

Radio galaxy: A galaxy emitting vast quantities of radio waves, often from two giant magnetic clouds straddling the galaxy.

Radio telescope: An instrument that detects radio waves from the cosmos.

Red dwarf: A small, cool, dim star.

Red giant: A large, luminous star with a low surface temperature and a reddish color. It burns helium in its core rather than hydrogen and is nearing the final stages of its life.

Redshift: The change in a galaxy's light as it recedes from us in the expanding universe, stretching its radiation toward longer, redder wavelengths — just as the pitch of an ambulance drops as it speeds away.

Reflector: A telescope that collects and focuses light with a mirror.

Refractor: A telescope that focuses light with a lens.

SETI: The Search for Extraterrestrial Intelligence.

Shooting star: See *meteor.*

Solar eclipse: When the Moon moves in front of the Sun, blocking off its light. In a total solar eclipse, the Moon covers the Sun entirely and we get to see the faint solar atmosphere, the corona.

Solar flare: An explosion that's created when the magnetic loops of the Sun touch above a sunspot group and short-circuit.

Solar storm: A solar flare or coronal mass ejection.

Special relativity: Einstein's theory of space, time, speed and energy.

Star: A sphere of glowing gas that generates energy by nuclear fusion in its core — like our Sun.

Stellar mass black hole: A lightweight black hole that results from a supernova explosion.

Steady state theory: A now discredited idea that the Universe has always existed and is changeless.

Supermassive black hole: A giant black hole that forms the heart of a galaxy.

Supernova: An exploding star.

Sunspot: A cooler and darker patch on the Sun's surface, where loops in the Sun's magnetic field restrict the motion of its gases.

Starburst galaxy: A galaxy suffering an outburst of starbirth at its center.

Transit: When a planet crosses the face of the Sun as seen from the Earth.

UFO (unidentified flying object): An object seen in the sky that's not immediately identifiable.

Van Allen belts: Two doughnut-shaped regions around the Earth where our planet's magnetic field traps very fast protons and electrons.

White dwarf: The fading core of a dying star that has lost its outer layers.

Wormhole: A potential shortcut through space, either across our universe or to another universe.

1 kilometer (km) = 0.62 miles
The international standard for astronomy is to use kilometers when referring to distances.

INDEX

Page numbers in *italics* refer to photographs.

INDEX

IMAGE CREDITS